IT'S A WRAP

CONVERSATIONS ON PLASTIC

Anne Hayden, MSc

Copyright © Anne Hayden (2023).

Anne Hayden asserts the moral right to be identified as the author of this work.

ISBN: 978-1-7396189-0-2

Cover design by Niall MacGiolla Bhuí and Dorothy Dryer
Published by ShadowScript Publications
An Independent Publishing House with Offices in Galway, Dublin, Ireland and Lund, Sweden

www.shadowscriptghostwriters.com

ShadowScript Publications is committed to inclusion and diversity. We print our books on forestry sustainable paper. Viva la forests!

Dedication

This book is dedicated to the loving memory of my late father, Patrick Hayden. His kindness, compassion, unwavering support and profound belief in me were the bedrock upon which this book and my career to date are built. As I ventured into the world of writing, it was his encouragement that inspired me to take on this challenge. Growing up with my father, with his deep-rooted connection to agriculture and his profound love and reverence for nature, left an indelible mark on my perspective on the world.

His forward-thinking mindset and his unwavering belief in the interconnectedness of the environment have been a constant source of inspiration throughout my journey in crafting this book.

Though he may no longer be with us in person, his legacy lives on in the pages of this book, a testament to his enduring influence and the profound impact he had on my life. As we face the formidable challenges that lie ahead, I am reminded of the words of my late father: "The impossible just takes that little bit longer."

Patrick Hayden, a wonderful man and a beloved father, this work stands as a tribute to you, with my deepest gratitude and love now and always.

Acknowledgements

As the saying goes, "It takes a village to raise a child" and this book is no exception. My heart brims with profound gratitude for the exceptional community at ShadowScript Publications, who selflessly devoted their time and efforts to breathe life into my vision.

To the remarkable contributing authors in this book—Dr. Niall MacGiolla Bhuí, Dr. Phil Noone, Cathy Fitzgibbon, known as The Culinary Celt, and Conor MacGiolla Bhuí—I extend my deepest appreciation. Your voices have lent depth and richness to these pages, transforming this book into a harmonious body of work.

I would like to express my deepest gratitude to the cornerstone of my life, my beloved family. Their unwavering presence and support have been the basis upon which I've built this creative journey, and for that, I am profoundly grateful.

To my remarkable mother, Nora, and the memory of my beloved late father, Patrick, I owe a debt of gratitude that words alone cannot fully convey. Their patience and enduring encouragement have been the guiding lights that illuminated the path of this process. In my mother, I found strength and resilience, a fountain of love and support that has been an unshakable source of inspiration during this difficult time, for which I can never repay.

My deepest thanks to my lovely brothers Thomas and Donald and sister-in-law Miriam for their boundless kindness and unfaltering support throughout this entire journey. Their presence has been a testament to the warmth and love that permeates our family bonds. To my wonderful nieces Katie and Emma, your love and delightful sense of humour will always hold a special place in my heart. You both bring so much joy and laughter into our lives, and your unwavering encouragement means the world to me.

I want to extend my heartfelt thanks to all my friends for their unwavering support throughout the creation of this book.

My unwavering gratitude to my cherished adoptive Galway family – Niall, Susan, Conor, and Ruby. Your kindness and friendship throughout this entire journey have not only touched my heart, but have also left an indelible mark on my soul. In life, it is rare and precious to find individuals who not only open their arms but also their hearts, embracing you as one of their own, and I will be forever grateful for the kindness and support you have shown me.

Last but by no means least, I want to extend my heartfelt gratitude to my dear friend and esteemed colleague, Dr. Niall MacGiolla Bhuí. His unwavering encouragement and exceptional editing prowess have been the catalysts that transformed this ambitious endeavour into a reality over the past twelve months. Without his tireless support, none of this would have been possible, and for that, I am profoundly thankful. Niall's dedication to this project has been nothing short of remarkable. His insightful guidance, meticulous editing, and unwavering belief in the vision behind this work have not only enhanced its quality but also pushed me to reach new heights as a writer. His contributions have been invaluable, and his friendship is an enduring source of inspiration. It is a privilege to have such a dedicated colleague and friend as an editor, and I am deeply appreciative of the role he has played in bringing this project to fruition.

Foreword

This book on plastic and its effects on us and the world we live in is a topic that I am extremely passionate about. I wanted to produce a body of work that explores the complex connection between plastic and its significant effects on our daily lives and the world at large. It is a topic that has captivated my interest and sparked a strong sense of concern. During my academic pursuits, I have been struck by the contradictory casual attitude society exhibits towards plastic, despite its wide-ranging implications.

Plastic, which is widely used in our modern society for its versatility and convenience, has become a constant presence in all aspects of our lives. From packaging to transportation to healthcare, it seems to be everywhere we turn. However, this widespread use comes at a cost - the alarming issue of plastic pollution that continues to escalate with each passing day.

As I embarked on my academic journey, I couldn't help but question the choices we make as a society. Why do we persist in selecting plastic for its convenience despite the undeniable risks it presents to our environment, wildlife, and, ultimately, our health? This curiosity ignited a fervent desire to deepen my understanding of this subject. With every page of this book, I aim to unravel the complex web of plastic's influence, from its origins and manufacturing processes, to its lasting imprint on the environment. But my passion extends beyond the scientific facts and figures. I want to ignite a dialogue, a collective awakening, and inspire action. In the following chapters, we examine the extensive history of plastic and its diverse applications. We will also delve into alarming statistics related to plastic pollution and explore promising solutions that are currently being developed. Through this exploration, we aim to gain a comprehensive understanding of the impact of plastic on our ecosystems, human health, and the global environment.

Plastic is not perceived as a glamorous theme to explore. I have infused the start of each chapter with a sound dose of millennial humour, taking readers on a jaunt of popular culture references. The purpose behind this approach is to enhance accessibility for all readers and should not be misconstrued as undermining the importance of this subject.

My interest in plastics and their utilisation began during my final year thesis for my undergraduate degree in Food and Agribusiness Management at University College Dublin, Ireland. The focus of my research was on consumer and industry perceptions regarding plastic food wrapping and its recyclability, with the results published in An Appraisal of Attitudes to the Environmental and Chemical Hazards of Plastic Food Packaging (2022). While studying at Imperial College, our group explored the potential use of edible plastic packaging as a promising solution to this escalating issue. However, we have yet to witness significant growth in this industry as expected. As part of my master's program from the Smurfit Business School in Dublin, I learned of the crucial role that the packing of a product plays in affecting marketing strategies for various campaigns. Yet, despite these revelations, one sobering reality remains: the broader public still lags in comprehending the paramount significance and far-reaching impact of plastics in their daily lives.

From my vantage point at my office desk, overlooking the bustling Galway docks, I am witness to a weekly spectacle that never fails to leave me fascinated. Right before my eyes, enormous ships are laden with bales of plastic, destined for foreign shores (in this week's case, Panama) to either meet their end in disposal in incineration or landfill or, we hope, embark on the intricate journey of recycling. It is in these contemplative moments that a thought invariably takes hold: should we not aspire to do better, and do we not, as a society, deserve better?

—Anne Hayden, September 30th 2023

The Urgency of an Open Dialogue: The Significance of Discussing Global Plastic Consumption

---

Niall MacGiolla Bhuí, PhD Director, ShadowScript Ghostwriters

In a world where plastic waste threatens our environment and endangers countless species, it is now crucial that we engage in a public discourse on global plastic use. Through exploring the economic, political, and cultural factors that contribute to this issue, we can develop more comprehensive solutions that protect both our planet and local, regional, national and international communities.

In this important book, Anne Hayden argues for the significance of having a transparent dialogue on plastic usage at all levels of society in order to facilitate meaningful change towards real and long-term sustainability. Anne has travelled to several European countries and accessed countless journal articles and books to gather comparative data on the theme of global plastic (mis)use. This is, after all, our one and only precious world!

So, plastic is ubiquitous in our daily lives. From the food packaging we encounter at practically every supermarket around Europe (as evidenced by the photography of our international research associate, Conor in this publication) to the water bottles we carry on hikes and onto sports fields and tracks, plastic has become a staple of what is understood to be a modern convenience. However, this convenience comes at a cost: every year, millions of tons of plastic waste end up in our oceans and landfills, with very serious consequences for both human health and the environment, as Anne explains across her chapters. And, we are learning daily what some of these consequences are.

The scale of this issue is staggering - it's estimated that by 2050 there will be more plastic than fish in the ocean. Just think about that for a moment! Many times I have been sea swimming off Blackrock in Salthill in the cold Atlantic Ocean, only to have discarded bottles, wrapping and caps as unwelcome travellers. In recent years, there has been growing awareness about the environmental impact of single-use plastics and calls for action to address this craziness. While governments have taken (some) steps towards reducing plastic waste through legislation including bans on certain types of plastics or imposing fees on single-use items like bags or straws, repeated individual actions are also necessary to make meaningful progress.

This brings us to an important question: how can we, as individuals, take effective action against plastic pollution? How can we open up and engage in conversations on plastic? The answer lies not only in individual choices but also in collective efforts towards sustainable solutions. It requires education and awareness-raising among individuals from all walks of life so they better understand their roles in contributing to this problem and, of course, to solutions. We can decide what changes we need to make personally.

I've written quite a lot on social media use amongst Millennials and Gen Z and I believe that discussing global plastic usage publicly through various media channels can play a critical role in raising awareness about its harmful effects on our environment while promoting engagement towards finding sustainable alternatives. And this is key. People must come together as communities worldwide—not just in a local or regional context—to share workable and realistic ideas for reducing waste production. This cannot be focussed solely around plastics consumption alone but speaks to broader systemic approaches focused around circular economy models which help reward innovation within industry practices ultimately addressing climate change more broadly too!

Let's look briefly at some research. According to McDermott (2016), an alarming 8 million tons of plastic enter the oceans annually, with 80% of this being single-use plastic. This horrific statistic alone highlights the magnitude of the problem that we are facing today. More recently, Teh et al. (2022) emphasise that plastic pollution is such a significant environmental issue that it requires immediate attention from individuals, organisations and governments alike. The authors report on an experiment in which they collected over 274,000 tweets related to plastic pollution and applied topic modelling techniques to understand public opinions about the matter on social media platforms. Their study found that popular topic modelling techniques such as Latent Dirichlet Allocation (LDA), Hierarchical Dirichlet Process (HDP), Latent Semantic Indexing (LSI), Non-Negative Matrix Factorisation (NMF), and Structural Topic Model (STM) were effective in detecting and identifying popular topics surrounding plastic pollution.

This finding by Teh et al. suggests that there is potential for developing more efficient systems aimed at addressing global plastic waste management issues collectively through social media platforms. According to World Economic Forum's White Paper: Plastics, Circular Economy and Global Trade Executive Summary ("World Economic Forum," 2020), E-Trade policies could be one solution towards addressing these issues effectively by encouraging sustainable solutions towards reducing wastes or increasing recycling rates among countries across different regions globally.

Thus, having a public conversation regarding global use of plastics is crucial if we want to address this issue successfully since individual actions can only go so far without collective action from all stakeholders involved—the government, industries, communities worldwide need awareness-raising campaigns aimed at promoting sustainability practices better aligned with circular economy principles like reusing rather than disposing entirely used products into landfills or marine habitats where they cause severe harm both environmentally but also economically because of the loss of resources and potential benefits from recycling or repurposing them.

The bottom dollar, pound, euro, yen or whatever your currency of choice is remains that plastic usage must be looked at much more rigorously than heretofore. To effectively tackle this issue, there is a need for a public conversation that educates people on the effects of plastic use on the environment and on us as human beings. This conversation should encourage individuals to take collective action towards sustainable solutions and it should be respectful of people's access to money and resources to effect change. It is not enough just to be aware of the negative effects of using plastics; we must also be motivated to take practical steps towards achieving sustainability and extending our hands to people less well off financially who simply cannot look ahead as they are concentrating on survival in their own communities and not on potential plastic use hazards. We've all seen the footage of patronising leaders flying in their private jets to sustainability conferences. How unbelievably annoying! Having open discussions will increase awareness about environmental issues associated with plastics use while providing an opportunity for diverse perspectives and innovative ideas. And this diversity must be appreciated.

Public conversations can prompt governments, industries, companies and individual consumers into changing their production or consumption patterns concerning single-use plastics.

Creating such forums helps build community cohesion by bringing together different stakeholders, at different levels and strata, who share common interests in preserving natural resources against human activities' adverse effects like inappropriate disposal of plastic wastes.

Future research could explore how social media platforms can provide public spaces where these conversations may occur on various levels - local communities around specific topics – globally through hashtags dedicated specifically addressing this subject that promotes online activism through posts sharing information plus calls-to-action encouraging users' engagement with relevant campaigns targeted at reducing or ceasing reliance upon non-biodegradable materials in everyday life.

The excessive and irresponsible use of plastics has caused severe damage to our environment, including marine life and human health. Therefore, it is imperative that we address this issue as a society. Individual citizens can make small changes in their daily lives to reduce plastic consumption. However, these individual actions alone are not enough. It should be emphasised that addressing global plastic use is not only an environmental concern but also a social justice issue.

The communities most affected by plastic pollution are low-income and marginalised populations who lack access to clean water or proper waste management systems. We've all seen the horrific photographs of plastic being washed up on Caribbean shores where people are knee deep in plastic flotsam and not clear blue water! People's thoughtless behaviour on one side of the world negatively affecting people thousands of miles away does not seem fair.

Therefore, advocating for systemic change at all levels is the way forward. Only then can we create a healthier planet for ourselves and future generations. Only then can we improve our own health and those of our planetary neighbours. All species.

Works Cited

McDermott, Kristin L., 2016, "Plastic Pollution and the Global Throwaway Culture: Environmental Injustices of Single-use Plastic". https://digitalcommons.salve.edu/cgi/viewcontent.cgi?article=1001&context=env434_justice

Teh, P.L.; Piao, S.; Almansour, M.; Ong, H.F.; Ahad, A., 2022, "Analysis of Popular Social Media Topics Regarding Plastic Pollution", MDPI. https://eprints.lancs.ac.uk/id/eprint/163469/1/sustainability_14_01709.pdf

World Economic Forum, 2020, "White PaperPlastics, the Circular Economyand Global Trade". https://www3.weforum.org/docs/WEF_Plastics_the_Circular_Economy_and_Global_Trade_2020.pdf

Other Sources

Pew Charitable Trusts, "Breaking the Plastic Wave". https://www.pewtrusts.org/-/media/assets/2020/10/breakingtheplasticwave_distilledreport.pdf

Baillie, Jonathan, 2018, National Geographic Society. https://www.epw.senate.gov/public/_cache/files/7/e/7e7dd259-af4c-475d-856b-989846b6aebd/D2CF0467AE2DF2F03D32F06DF246CEE3.baillie-testimony-09.26.2018.pdf

Table of Contents

Foreword i

The Urgency of an Open Dialogue: The Significance of Discussing Global Plastic Consumption iv

Chapter 1 The Promise of Plastic: The Rise, Hype, and the Epic Fail – A Hugely Promising Start and Its Fall from Grace 1

Chapter 2 Unwrapping The Truth: The Jackal and Hyde Paradox of Plastic on Food Safety 22

Chapter 3 Making a Splash: Plastics in Our Oceans 56

Conclusion 79

GUEST CONTRIBUTORS

The Green Revolution 85

Forty Shades – A Green Revolution Blooms! 87

The Notion of an Ocean 101

Plastic Use and the Need to Protect our Oceans 103

The Future is in Our Hands- A Voice From Gen Z 117

The Global Plastic Waste Crisis Through the Lens of Gen Z: Exploring Perspectives and Innovating Solutions 119

About the Author 131

Chapter 1
The Promise of Plastic:
The Rise, Hype, and the Epic Fail – A Hugely Promising Start and its Fall from Grace

"I Came in Like a Wrecking Ball."
(Miley Cyrus, 2013)

When examining causation regarding events or product failures, it is crucial to retrace our steps back to the very beginning. These incidents are often not isolated occurrences that lead to systemic failures. Plastic products, for example, were initially hailed as remarkable innovations, but their true impact often remains concealed until it is too late. The consequences of plastics can be likened to Miley Cyrus's iconic lyrics: "I came in like a wrecking ball, I never hit so hard". It caught us all off guard until the unforeseen repercussions of plastic became an undeniable reality. Just like Miley's wrecking ball left behind devastation, the effects of plastics have sent shockwaves throughout our environment, ecosystems, and communities; leaving us grappling with the unanticipated outcomes of this once-promising wonder product.

When delving into the intricacies of complex issues, I've discovered that drawing parallels to historical events helps my neurodiverse mind grasp these multifaceted concepts. In my quest to find an analogy to explain the profound impact of plastic on our environment, ecosystems, and our daily existence, the Titanic emerges as a poignant comparison. The rise and fall of plastic shares striking similarities with the tragic saga of the Titanic. Both narratives share a common trajectory that begins with immense hope, promise, and investment but ultimately encounters an unanticipated crisis with a tragic end. Similar to how the Titanic collided with an iceberg beneath the surface, plastic's pervasive influence has resulted in an environmental and ecological catastrophe hiding below our awareness. The Titanic was hailed as an unsinkable marvel, a symbol of human progress, while plastic was championed as a versatile, modern material promising endless possibilities. The trajectory of plastic closely mirrors the dramatic story of the Titanic.

Both started with high expectations, receiving considerable support and investment from so many sources. The Titanic was esteemed as an unsinkable achievement, representing human advancement, just like plastic was embraced as a flexible and innovative material that held limitless potential.

Unfortunately, just like the Titanic's collision with an iceberg caused its tragic demise, plastic has contributed to a massive environmental crisis known as plastic pollution. This widespread issue is wreaking havoc on our oceans and ecosystems, turning what was once a hopeful material into a global growing concern. The parallels continue just as the Titanic's passengers in steerage faced the greatest peril. The environmental impact and health hazards associated with plastic disproportionately burden marginalised communities, echoing the class divide aboard the ill-fated ship. In both cases, the initial optimism and prosperity have given way to profound challenges, highlighting the need for responsible stewardship and a collective effort to address these crises.

Inside Out: Investigating the Complex Interaction Between Plastic and the Human Body

Plastic surrounds us in our lives; it is a crucial part of our daily routine. We rely heavily on plastic, from household materials and appliances to packaging materials for food and other products. However, as convenient as it may be, this material poses potential health risks to the human body. The relationship between plastic and human health is complex and multifaceted. While offering an essential component to our daily lives, plastic also seriously affects our health.

This chapter takes an in-depth look at how long-term exposure to various forms of plastics can affect human physiology and suggests ways to minimise the risk involved in having plastics embedded so deeply within our society. We discuss how the health implication of having so much plastic packaging and the chemicals that are used in this packaging affect our health.

Several chemicals used in plastic material, such as Bisphenol A, phthalates and flame retardants, affect the endocrine system negatively, and these will be discussed as these chemicals have been found to disrupt hormonal balance and interfere with reproductive development. Research shows that some developmental stages are more vulnerable to the toxic effects of chemicals associated with plastic waste.

Microplastics and their influence on human health are becoming increasingly important. Plastic microscopic particles have been discovered in a variety of sources, including food, water, and even air samples. As micro plastics are commonly ingested through a variety of sources according to studies, exposure to these microscopic pollutants may cause injury to human organs such as the liver, kidneys, and lungs, affecting general health. It has also been proposed that phthalates, which are commonly employed as plasticisers, contribute significantly to endocrine disruption, resulting in negative effects.

The detrimental impact of plastic on human health extends to its potential effect on the immune system. Certain plastic components can provoke immunological responses that prompt allergic reactions, autoimmune disorders and weakened immunity, among other risks. Researchers are currently studying how plastic interferes with the human microbiome - a complex bacterial ecosystem within our bodies - as concerns grow how they can trigger extensive effects on the overall well-being and health of both humans and sea creatures. Various studies have shown that prolonged exposure to plastics and their associated compounds can have major consequences for human physiology and general health, including human mental health, and these will also be discussed in this chapter.

Sources of Plastic Exposure

Plastic pollution is a pressing global issue that has severe implications for the environment and human well-being alike. Its rampant use in manufacturing, consumption practices, and improper disposal methods have led to its widespread distribution throughout various ecosystems worldwide - from oceans and rivers to forests - even finding its way into the very air we breathe. The sheer scale of plastic's presence in these environments poses significant risks to public health as it endangers our ecosystems' integrity by disrupting natural habitats vital for biodiversity upkeep. In order to address this growing problem, there is an urgent need to expand on ways we can mitigate or prevent plastic pollution effectively.

This involves incorporating interdisciplinary approaches from environmental science, and engineering technologies alongside improved environmental and health policy and intervention strategies designed to reduce plastic wastes at all stages of the production-consumption-disposal cycle.

One of the major issues that cause alarm is microplastic ingestion, which refers to the consumption of minuscule plastic particles measuring less than 5mm found in water resources and marine life. These fragments could accumulate within human systems, resulting in inflammation, organ injury, or malfunctioning of physiological processes. Besides this effect on humans' health when ingested blatantly, they also pose a danger as hazardous chemical agents such as persistent organic pollutants can adhere to their surfaces; thus contaminated plastic conveys risk-laden elements indirectly via mass exposure through microplastic intake by living organisms, including humans.

Microplastics, tiny pieces of plastic waste that have become pervasive in our environment, pose a grave threat to the marine ecosystem and human health. It was considered for several years, that microplastic pollutants found in seafood consumption were the main source of microplastic consumption, which might be ingested by humans, entering their bloodstream as potential health hazards. This endangers aquatic organisms that ingest them and has severe consequences on human health. However, it has been established that sea food consumption is in fact not the only way in which we are consuming microplastics. Plastic pollution in food and drinking water demands immediate attention as it poses significant risks to human health. Plastic containers, utensils, and bottles release microplastics and microplastics that contaminate the substances they carry. These minute plastic particles can enter human bodies through ingesting contaminated food or water; therefore, researchers have found trace amounts of microplastics in seafood, salt, and even drinking water just to name a few sources implying a widespread presence of this pollutant across multiple contexts.

The contamination of the environment with microplastics has been a growing problem for several years, resulting in studies aimed at its effects on marine water and ecosystems. Several studies have found them in a variety of water sources around the world, including rivers, lakes, and even groundwater reserves. However, recent research has also expanded to address drinking water, revealing worrying findings. Both tap and bottled water have been found to contain microplastics of varying shapes, sizes, amounts and plastic types. This alarming discovery prompted the World Health Organization to release reports focusing on drinking water quality and its impact on human health and to encourage continuous monitoring of the quality of our drinking water. Microplastics in water can come from a variety of sources, these particles can reach the water supply via a variety of means, including air deposition, surface runoff, and sewage treatment procedures.

The scientific evidence reveals that microplastics can travel through the atmosphere for both small and large distances because of atmospheric movement aided by wind patterns, usually for distances up to and exceeding 100km. The deposited particles of these plastics are subject to re-suspension, which eventually leads them back into the water column and increases their chances of entering aquatic ecosystems. Effectively, this means that microplastics are not only present in our oceans and in the food we consume, but Microplastics are now in the air we breathe; as humans, one of our most primal requirements for survival depends on oxygen in the air to exist for survival and proper organ function.

As a result, because plastics are prevalent in our environment, humans are breathing them in, and they are now present in our nostrils, mouth and lungs (1). Expanding on this premise of plastics now infiltrating our bodies, some suggest by scientists that future research needs to examine the link between microplastics and their ability to infiltrate pores on human skin (1). Plastic is not only present in our everyday lives it is literally becoming part of us and part of our bodies, the overall health effects that this has on us have not yet been examined, but the results to date have been startling and will be discussed further in this chapter.

One of the critical environmental concerns associated with water pollution is runoff. Rainfall or irrigation systems often carry large volumes of sediments and other pollutants from various sources and deposit them downstream, contaminating larger bodies of water. The contamination of surface runoff from roads, residential areas and fields is caused by several factors, including rainfall intensity, nutrient distribution practices, land use patterns and other pollution-causing hazards, and so it is complex when trying to decipher a main cause. Because of their high impermeability, scientists have discovered that main road surfaces are among the most significant microplastic contributors to the runoff water being contaminated with microplastics. Research has shown that these routes frequently experience higher traffic volume concentrations leading to increased levels of plastic pollution from wear and tear of tire (tire abrasion) on the road and discarded rubbish which is not absorbed by the road or motorway, thus adversely affecting the entire ecosystem for aquatic life (2).

As runoff acts as a direct channel for water pollution, carrying large quantities of accumulated materials downstream where it gathers, permeating surface water, rivers, lakes and stream, creating an adverse effect on marine environmental biodiversity preservation efforts. In contrast, runoff from fields is now less of an issue because of the development and implementation of EU directives such as S.I. No. 113/2022 - European Union (Good Agricultural Practice for Protection of Waters) Regulations 2022, preventing water contamination from sprays and fertilisers through correct usage and monitoring. We need to implement this monitoring and evaluation of our waterways from not just pollution but from microplastic in order to address this growing concern. Implementing better storm management practices which would allow for the run-off water to gather in specific areas, which then can be filtered to remove these pollutants and plastic particles before being allowed to join the waterways (lakes, ponds, rivers etc.), would help in reducing the pollution of the waterways.

Ineffective sewage treatment procedures also cause microplastics in bodies of water. Sewage treatment processes, although reliable at removing solid particles and organic materials from wastewater, contribute significantly to water's microplastic pollution. While the process used for filtration and sedimentation is equipped to deal with the removal of many types of contamination levels of nutrients such as phosphorous and nitrogen, the small size and varied types of microplastics present make efficient removal challenging for wastewater treatment plants. Microplastics that escape the wastewater treatment facility can endure in the liquid wastewater and ultimately may be released into adjacent bodies of water such as rivers, lakes, and coastal regions. While this water is now technically treated and safe, the accumulating amounts of microplastics present have not been addressed. This water, if it enters bodies of water such as rivers and lakes, can affect aquatic life and can make its way back into our food chain. Another use of this water can be for agricultural irrigation, which may introduce microplastics into soil, which may eventually reach crops' systems, causing further dangers to public health through food consumption pathways.

Pathways of Plastic into the Human Body

Plastic pollution has now become a pressing concern as various studies have indicated the disastrous impact it poses on ecosystems, marine life and now on our own health. In addition to this, the production of plastics from petroleum-based materials is causing environmental pollution, leading to a depletion of fossil fuel resources; these concerns add further complexity, showing how closely microplastic issues are coupled with broader environmental problems. Microplastics can be traced back to a variety of sources.

One significant source is the degradation and fragmentation of bigger plastic products, such as packaging materials, bags, or bottles, caused by mechanical stress, weathering, or UV radiation. Over time, these microplastics undergo physical disintegration processes, resulting in tiny particles known as microplastics. As the interest in microplastics grows, it has become apparent that deliberate measures are taken to manufacture them for various purposes. Examples include producing microbeads found in beauty and personal hygiene products or synthetically creating microfibers used in textiles. Unfortunately, these deliberate types of microplastics have detrimental effects on the environment as they are released into water systems during usage and washing processes and help to continue this perpetual cycle we find ourselves in with plastics in our oceans now becoming part of not only our daily lives but also part of our physical makeup. The production and use of household items and plastics available in markets are additional sources that contribute to microplastic consumption by humans.

According to recent scientific research, microplastics - an emerging pollutant - are present in the air we breathe, the food we eat and the water that we drink. Humans are exposed to these harmful particles via multiple routes which infiltrate our bodies, giving rise to severe health implications. These findings bring forth substantial worries over the many dangers associated with contamination by microplastics within our systems because of their prevalence and omnipresence across different environments such as drinking water supply, and freshwater sources like rivers where wastewater treatment plants deposit them inefficiently contaminating river waters used for public consumption- all of this putting human lives at risk. The ingestion of microplastics is a major source of concern, as these tiny plastic particles can accumulate within the human body and potentially cause inflammation, organ injury, and reproductive issues.

The presence of microplastics in our daily lives, which we may unknowingly be exposed to, is a significant concern. However, it's important that we acknowledge and take ownership of the excessive use of plastic products in our lifestyles as well. In particular, research has raised concerns about single-use plastic water bottles because of their potential impact on human health through the leaching of microplastics from these containers into drinking water. Polyethylene terephthalate (PET) is commonly used in the production of plastic water bottles. Concerns have been raised over its potentially hazardous impact on human health when exposed to high temperatures or prolonged usage.

Recent research has showed that bisphenol A and phthalates can escape into the water contained within these containers as they gradually degrade with use, which could pose serious long-term risks to individuals who consume such contaminated drinking water regularly. These groups of chemicals, referred to as endocrine disruptors, have been associated with several detrimental health effects, including but not limited to disturbances in reproductive system functions, anomalies during development, and imbalances in hormonal regulation. Endocrine disruptors, which are chemicals that interfere with the body's hormonal system and have been linked to various adverse health effects, are found in plastic water bottles. The leaching of these harmful substances from plastic into our environment and bodies is a growing concern for human well-being. It highlights the necessity of finding safe and sustainable alternatives to disposable plastic water bottles, as their long-term impact on human health is yet unknown. Advancements in research surrounding endocrine disruptors show the need for continued research in this field towards ensuring optimal public policies promoting environmental preservation efforts while prioritising the population's protection against hazardous substances such as plastics contaminants.

Microplastics have been found to act as carriers of toxic chemicals like persistent organic pollutants that adhere to their surface. This not only elevates the risk of plastic contamination indirectly through mass exposure via microplastic intake by living organisms, including humans, but it also raises concerns about the long-term consequences of plastic pollution on human health and ecosystem sustainability. The mounting evidence of the hazards associated with microplastic pollution highlights the urgent need for coordinated efforts-from governments, industries, and individuals-to reduce plastic waste and prevent its release into the environment.

It has been showed that these micro plastics of less than 5mm in size can enter human bodies through various pathways of exposure such as contaminated food or drinking water. Recent empirical data suggests two possible modes by which these minute-sized debris can infiltrate deep into tissues

- first via immune cells through the lymphatic system and second, directly from ingested sources crossing over intestinal barriers leading to translocation across blood vessels. There is a rising concern and investigation into the existence of microplastics in the human blood. Although it has not been fully established yet, research proposes that there are other means through which these synthetic particles occur within organs; for example, via migration through the lymphatic system or transfer from the gastrointestinal tract to the systemic circulation. The movement of microplastics can be facilitated by immune cells moving them along lymph vessels to different tissues while they are ingested orally, crossing intestinal barriers and accumulating in various body parts. This highlights the urgent need for more comprehensive research to be conducted on the potential effects of microplastics on human health and their contribution towards toxic accumulation in the body.

Plastic, but more specifically microplastics, has become part of our body composition and needs to be urgently addressed.

Clearly, microplastics in our drinking water raise serious concerns about human health. The widespread use and production of plastics has led to increased exposure to these particles, with ingestion being the primary route of entry into the body. While more research is needed on the potential long-term effects of microplastics on human health, urgent action must be taken to reduce plastic pollution and limit exposure for individuals. In addition to individual efforts such as reducing plastic consumption and waste, policy changes are necessary at a larger scale to address this global threat and protect public health.

Plastic's Impact on Human Health

Concerns regarding the influence of plastic on human health have grown in recent years. Plastic is abundant in our daily lives, from food packaging to personal hygiene goods; not all plastics are necessary for modern living because some may contain harmful substances that might infiltrate into people's bodies or the environment. In this chapter, we will look at the potential health hazards linked with plastic exposure and how we may protect ourselves and future generations from its detrimental consequences. Recent research has identified various exposure routes through which microplastics can make their way into human tissue, such as lung and placental tissues. The toxic effects of these tiny particles have also been documented with reports suggesting that they trigger inflammation, oxidative stress or even cell death.

Studies show that certain chemicals commonly found within plastic materials - including Bisphenol A, phthalates and flame retardants - wreak havoc on endocrine systems in humans and animals alike

The extent of risk posed to humans, marine life and the ecosystem by plastic pollution is multifaceted, given that each type of plastic has unique properties that impart different effects on ecosystems. Hazardous chemicals contained within plastics are among the major worries associated with such pollution because of their potential for permeating our surroundings or bodies. Bisphenol A, phthalates, and flame retardants present in many kinds of plastics have been linked as causative of unfavourable health outcomes; therefore, recognising possible risks alongside taking protective measures is crucial toward safeguarding both current and succeeding generations against exposure from ingestion of contaminated food or water sourced from degraded plastics or inhaling dust particles from them. Research has shown that microplastics found in the environment can also have a negative impact on our gut health and cognitive function.

Bisphenol A is a widely utilised chemical molecule that is employed in the production of various plastics and resins. It can be detected in an assortment of consumer goods, such as containers for food and drinks, dental sealants, and thermal paper receipts, among others. Bisphenol A is more commonly known by the name BPA. because of its endocrine-disrupting properties - which imply mimicking or interfering with hormone action within the body- BPA has been linked to many health issues, including disruptions in hormonal balance, reproductive abnormality, developmental disorders; and augmented risk of cancer according to diverse scholarly undertakings conducted on this subject.

There are also studies showing associations between BPA exposure and cardiovascular diseases, obesity, diabetes as well as neurological concerns; because of these negative outcomes associated with BPA, many products have not had to become and be advertised as BPA-free. In order to mitigate the dangers of Bisphenol-A on human health, it is important to not only minimise exposure to BPA but also promote alternative materials that are free from its presence. Using safer options for all plastic products used by people should be encouraged. Originally, BPA was commonly used as a component in children's plastic cups and bottles, which unintentionally caused harmful effects. Various studies have showed indisputable evidence connecting children's exposure to BPA with neurobehavioral impairments, including attention-deficit and hyperactivity disorder (ADHD).

Expanding research on the potential dangers of Bisphenol-A causes efforts to minimise its use and advocate for alternative options that are safe from this chemical compound. It is imperative to raise awareness about how minimising human exposure can prevent the negative affects of Bisphenol-A on their health, particularly in developing infants or early childhood stages where sensitivity is at its highest point. This is currently just one type of plastic molecule derivative that causes these harmful effects, but as research is increasing into this area, other knock-on effects are also being observed.

Similarly, the use of phthalates in plastic production is of serious concern for human health. The application of phthalates as plasticisers in producing plastics raises a grave concern for human health. Phthalates belong to a group of chemical compounds commonly used to enhance plastic products' flexibility, resilience, and transparency. Various consumer items include but are not limited to children's toys, shower curtains, vinyl floorings, as well as some beauty cosmetics and personal care items. Phthalates can enter the body through ingestion, inhalation or skin contact, which raises serious concerns about their potential harmful effects on human wellbeing. As these additives seep into everyday objects that humans come in close contact with regularly, it is vital to recognise the sources of contamination caused by these chemicals for better protection against the risks they pose to people's health.

Phthalates, known for their potent endocrine-disrupting properties, have been linked to several adverse health effects (3). These chemicals can interfere with the hormonal system in humans and adversely impact reproductive function, which may lead to decreased fertility. Studies have documented that phthalates are harmful to male reproduction and can cause disruptions in the reproductive system (3). Therefore, it is essential that further research be conducted on these chemicals not only for environmental purposes but also for human health concerns.

In addition to the hazards posed by plastic polymers, research has shown that certain chemicals used in plastics may also put consumers and workers at significant risk. Chemical additives such as phthalates have been associated with a broad range of health conditions, including asthma, allergies, obesity, and cardiovascular disease. In fact, recent studies suggest there is even more cause for concern for exposure amongst children, as this group may be vulnerable to adverse neurological effects arising from these compounds (3). Phthalates are known endocrine disruptors, which makes their effects on human reproductive function and hormone balance, particularly concerning, especially as children are being exposed to these disruptors from an extremely young age, which may have unforeseen consequences on their health as they age (4).

Similarly, the use of flame retardants such as polybrominated diphenyl ethers has been restricted or banned in many parts of the world because of their hazardous effects on human health. These chemicals pose a significant risk, especially when they leach into the environment from products like furniture and electronic devices. Some studies have linked PBDE exposure to developmental delays, reproductive disorders, and even cancer. Despite some positive actions taken by certain countries against these toxic substances, there's still much work that needs to be done regarding the global regulation of harmful chemicals used in consumer goods for increased public safety and improved environmental protection measures.

PBDE (polybrominated diphenyl ethers) has been extensively studied because of their potentially harmful impact on human health and the environment. Owing to these concerns, Europe has banned PBDE usage for manufacturing various commercial products, including furniture, electronics and automobiles. Despite this restriction in place since 2008 under REACH Regulation EC No 1907/2006 Annex XVII entry 67 issued by EU Commission regulation no 850/2014. However, former production capacity ensures that it may continue entering supply chain goods sold online without physical sales points control.

Many older items manufactured before stricter legislative action on hazardous substances were enacted pose significant risks during use or disposal because of leaching into soil or water systems through degradation processes like incineration or burial in landfills. Therefore, it is imperative for consumers to exercise caution when purchasing second-hand consumer commodities and check origin dates as well as compliance indications, even if these fall outside European limits. As we strive for increased recycling efforts to limit plastic waste production while reducing our carbon footprint from excessive consumption activities that might cause harm to wildlife— careful actions must be taken when purchasing second-hand consumer commodities by checking their origin dates carefully—even if they fall outside EU limits. Retailers should also shift their positioning towards less toxic alternatives such as phosphorus-based fire-retardant chemicals.

Therefore, correctly labelling plastics is essential for proper identification for safe removal from the market to prevent further distribution instead of possible reuse if not appropriate. Though flame-retardant serves its purpose when used judiciously, providing safety against fires; there are concerns regarding toxicity concerning ecosystems' stability, posing the question of whether the benefits outweigh the costs. A suggestion is that individual product life cycle analysis should be looked at independently by these companies, making the information on the effects available, ensuring correct conclusions so as to give consumers more freedom of choice whether to engage with the affected products already on the market.

The impact of plastics on the human endocrine system and health is an increasingly concerning matter among experts worldwide. There is a correlation between human exposure to plastic and inflammation, as well as oxidative stress in the body. The incorporation of substances like phthalates and bisphenol A into plastics can incite inflammatory reactions while producing reactive oxygen species, which disrupt cell functions that create an imbalance between immune responses. Prolonged bouts of these symptoms result in chronic diseases such as respiratory issues, neurodegenerative conditions, and cardiovascular disorders. Finding solutions to ease these detrimental health effects along with safeguarding environmental safety requires examining the means by which plastic pollution initiates inflammation signalling pathways via oxidative mechanisms critical for future research endeavours.

Excessive plastic exposure has been linked to increased levels of oxidative stress, which occurs when the number of harmful molecules, called reactive oxygen species, surpasses the body's ability to defend itself. Using plastic additives can worsen this situation by activating enzymes that generate more reactive oxygen species and interfering with mitochondrial function through different pathways. This overproduction of these harmful molecules because of plastic exposure damages essential cellular components like proteins, lipids, and DNA, resulting in impaired cell functionality or tissue damage caused by oxidation reactions on vital biomolecules within them.

Studies suggest that exposure to plastic may cause chronic diseases such as cancer, diabetes, and neurological disorders. High levels of lipid peroxidation caused by reactive oxygen species during increased oxidative challenge are associated with these health issues. However, the full extent of the damage is not yet fully understood.

Recent investigations suggest a fascinating association between gut bacteria and mental health. The gut microbiome, which consists of an enormous assembly of microbes (100 trillion microorganisms) residing in our gastrointestinal system, plays a vital role in various aspects of our well-being, including our psychological state (5). Evidence suggests that imbalances within the microbial population can trigger mental disorders such as anxiety and depression and a range of mood disorders (5). Via the bidirectional pathway connecting the brain with the intestine known as the gut-brain axis; microorganisms affect human behaviour and cognitive processes, influencing emotions while managing stress responses through the synthesis of neurotransmitters and changes to our immune functions, among other complex pathways. Discovering this correlation provides an interesting predicament what effect these microplastics that we are ingesting have on our gut health but also on our cognitive function and mental health (6).

CONVERSATIONS ON PLASTIC

The issue of plastics and their effects on human health are multifaceted and complex. While there is accumulating scientific evidence to suggest that plastics have negative effects on ecosystems as well as human health, particularly with chemicals such as Bisphenol A, phthalates, and flame retardants. It's clear we need to promote safe removal from markets or shop for less toxic alternatives like phosphorus-based fire-retardant chemicals by companies strictly adhering to the guidelines provided.

There also needs to be transparency in product labelling so consumers know whether a product contains hazardous substances they don't want to expose themselves to, as microplastic ingestion has been associated with gut health but also with mental health, making it crucial to investigate this correlation further and to stem any more negative effects from this. The concerns about adverse longer-term effects on human health and animal health highlight the need for more serious efforts towards awareness-raising both individually but mainly enactment of extensive measures, including usage reduction campaigns combined with improvements concerning disposal methods relevantly creating safer alternatives needed for environmental protection coupled with protecting human health and the environment.

Chapter 2
Unwrapping The Truth:
The Jackal and Hyde Paradox of Plastic on Food Safety

"I think I've seen this film before, and I didn't like the ending."
(Exile, Taylor Swift, 2020)

The use of plastic packaging in the food industry has become indispensable, providing convenience and protection for a wide range of products. However, the implications of this widespread practice on food security are still being inspected. In this chapter, we delve into the effects of plastic food packaging on food safety, including issues like chemical leaching, microbial contamination, concerns about consumer health, and its environmental footprint.

As research delves deeper into the realm of plastic food packaging, we witness a concurrent rise in the rigour of regulatory measures. Recent research has revealed that certain properties of plastics may pose potential health risks, leading us to reconsider our approach. With these findings in mind, it may be necessary to prioritise alternative methods of containing, wrapping, and storing food over traditional plastic options. The intersection of health and environmental concerns calls for a revaluation of our choices.

Understanding these complexities is essential for making informed decisions regarding the protection of consumable products and adopting environmentally responsible packaging choices. In modern manufacturing, there's been a growing reliance on plastic because of its cost-effectiveness and its ability to reduce fuel consumption by making vehicles lighter. Disposable items like fast-food packaging and picnic tableware are raising significant environmental concerns, contributing to a yearly waste accumulation exceeding one million tons.

While plastic food wrapping isn't entirely without merit, as we delve deeper into its implications and witness the emergence of stricter regulations, it prompts us to ponder whether it truly represents the safest approach for preserving and transporting food.

Let's once again explore a gradual shift that eventually resulted in its ban because of serious health concerns. Take, for example, the widespread use of asbestos—an industry that flourished primarily from the 1930s onwards as it became a popular choice for construction materials because of its excellent insulation and impressive fire resistance qualities. However, as we entered the 1950s and 1960s, concerns regarding the harmful effects of asbestos exposure started emerging. Extensive studies during this time established a clear link between asbestos exposure and debilitating lung diseases, such as asbestosis and lung cancer.

During the 1970s, governments in many countries, including the United States, began implementing stricter regulations and restrictions on asbestos use. Although there was a decline in its usage, it wasn't until 2000 that Ireland officially banned asbestos from being used in construction materials. Once more, we find ourselves grappling with a familiar quandary in the realm of plastic policy, where regulatory action often trails the unfolding consequences, emerging only after the issue has revealed its full impact. To borrow a sentiment from Taylor Swift's "Exile," it feels like we've witnessed this narrative before, and the ending was far from desirable.

The pressing question arises: Why do we delay making crucial policy changes, particularly when it comes to the use of plastics in our food products, despite mounting evidence of the associated health risks? This dilemma presents us with a weighty ethical inquiry: Should we not be compelled to do more and act with greater urgency as evidence continues to mount and not wait until it's too late to act?

Background to Plastic Food Wrapping

The plastic food wrapping industry is currently experiencing rapid growth, both within Ireland and on a global scale. This industry predominantly relies on low-density polyethylene (LDPE) for its products. While plastic food wrapping offers several distinct advantages, including tamper resistance, preservation of freshness, enhanced marketing capabilities, and prolonged shelf life for food items, it is imperative to recognise and address potential downsides, notably the risk of food contamination because of additives in plastic packaging materials.

The escalating issue of food packaging waste cannot be tackled in isolation; it necessitates a holistic approach, taking into account the rising use of plastic food packaging (6). While protecting food through effective packaging remains undeniably vital, it is now more critical than ever to strike a careful balance between preserving food quality and addressing the growing environmental and chemical concerns associated with plastic food packaging (6).

In recent years, there has been a significant surge in the widespread use of plastic food packaging, intensifying the complexities related to waste management (6). While there is no denying the crucial role that food packaging plays in safeguarding food quality, ensuring secure transportation, prolonging product freshness, and shielding against contamination (7), it's imperative to highlight that health-related concerns associated with plastic food wrapping continue to persist, even as its prevalence continues to grow (8).

Plastic food packaging consists of LDPE and is commonly referred to as wrapping film (8). The global LDPR market was estimated to be worth $46.65 billion in 2022 and is expected to reach $51.11 billion in 2023 (9). Food products deteriorate from the farm to the consumer because of the physical, chemical and biological changes (Han, 2014). The quality and safety of food products has improved because of food packaging and their extended shelf life (Han, 2014).

Food packaging is a marketing tool for food products as it benefits marketing with improved packaging and design (10).

Marketing increases sales as consumers are attracted to the products that have labels containing details about the product source, ingredients, expiry date, nutritional value and net weight (8). Plastic food packaging has advantages for the food producers as products can be sold all over Ireland, Europe and the world by extending the shelf life and maintaining the freshness of the food products (11).

Plastic food packaging has environmental and chemical hazards. Low-density polyethylene (LDPE) is widely used in food packaging as it is a low-cost material and has many advantages that benefit the suppliers and consumers of food such as different sizes and shapes, products are visible and are hygienically protected (8). LDPE is derived from the "chemical process for developed synthetic polymers (plastics)" (12). LDPE is manufactured by the poly addition of monomer units, where molecules link with double or treble carbon bonds to form a large molecule.

LDPE is mass-produced all over the world as it is easy to seal, strong, transparent and moisture resistant (8). The food producing companies use this plastic for food wrapping when supplying products to the retail outlets to such as chicken, apples, ham, salads, beef and bacon.

Plastics play a pivotal role in our daily lives, and understanding their composition is key to making informed choices. Each type of plastic is categorised by a resin code, and one of the most commonly used materials for food packaging is denoted by the #4 resin code, also known as LDPE (Low-Density Polyethylene) (8). These resin codes, ranging from #1 to #7, serve a vital purpose in facilitating the recycling process by necessitating the separation of different resin types (8).

CONVERSATIONS ON PLASTIC

While plastic food packaging undeniably offers numerous advantages, it's crucial to acknowledge the emerging concerns associated with its use. Increasingly, attention is being drawn to the potential negative effects on both consumers and the environment. Additives present in plastic food packaging have the potential to be absorbed into the food, raising health and safety concerns (8).

Beyond the immediate health implications, the environmental footprint of plastic food packaging is a growing concern. The manufacturing process generates harmful by-products and emissions, contributing to environmental degradation (13). The disposal of food packaging waste poses a significant challenge, as it adds to the mounting problem of plastic pollution. So, while plastic food packaging undoubtedly offers convenience, we must be mindful of its potential drawbacks. By understanding resin codes and the environmental implications, we can make more informed choices to balance the benefits and disadvantages of plastic packaging in the realm of food products.

Many studies have been conducted to explore alternative materials to replace plastics while ensuring durability across all applications, such as utensils or film wraps used in contact with food. These alternatives must not compromise the flavour or physical properties of edible items, making the search for sustainable solutions a critical aspect of our discussion. In the quest for eco-friendly substitutes, researchers are investigating materials like biodegradable plastics, plant-based polymers, and innovative composite materials. These endeavours are driven by the urgent need to reduce the environmental impact of plastic packaging without sacrificing the quality and safety of the products we consume. The pursuit of these sustainable alternatives represents a promising path towards a more environmentally responsible and consumer-friendly future.

Chemical Leaching from Plastic Food Wrapping

Despite being a widely used and practical solution for food preservation, plastic packaging has caused alarm regarding the potential chemical contamination of packaged food. Compounds found in plastics like phthalates, bisphenol A, and polybrominated diphenyl ethers can penetrate into consumable items and present likely health hazards for consumers. Many factors contribute to chemical leaching, such as temperature exposure incurred during microwave cooking or contact with hot liquids, possible storage duration period or composition of the substance stored that promotes migration into food content from its containers; acidic or fatty foods enhance this process even more so, intensifying their noxious effects on human health. It is essential that consumers are aware of these potential risks when utilising plastic food packaging and consider alternative options, such as biopolymer-based or biodegradable packaging, or switching to glass or metal containers for food storage and heating. Food manufacturers must take responsibility for ensuring that the packaging they use does not pose a threat to consumers' health.

What is becoming increasingly clear about chemical leaching from plastics into our food is how large the gap in our knowledge is in this area. While the use of plastic is not a new phenomenon, its understanding of its effects is new. The impact of plasticisers, such as bisphenols and phthalates, on human health is a topic that requires further academic exploration. Despite the widespread use of plastics, our understanding of their effects remains limited, and ongoing research is necessary to bridge this gap in knowledge. As these chemicals can easily leach from products into foodstuffs via different food-contact packaging materials during processing, storage, or transport, it's important for researchers and regulatory agencies to expand academically by conducting comprehensive studies to comprehend better the extent of chemical leaching processes on human health. In addition, alternative biopolymer-based packaging solutions offer an eco-friendly and non-toxic substitute, thus warranting additional application-focused studies around feasibility analysis across multiple scenarios before implementation at the retail level.

Furthermore, consumers need to be made aware of the potential risks associated with plastic food packaging and consider alternative options, such as glass or metal containers for storage and heating. It is incumbent upon manufacturers to ensure that their packaging does not pose a threat to public health by enforcing stringent quality control measures and adopting healthier composition practices. As the adverse health effects of chemical leaching from plastic food packaging are becoming increasingly clear, it is crucial for consumers to be made aware of potential risks from their use.

Several factors contribute to the migration of chemicals into packaged food from plastic wrapping. Temperature is a crucial determinant, as higher temperatures can speed up chemical transfer and intensify leaching. The type of food and its composition also play a role in this process, with certain chemicals being more prone to migration in the presence of specific components, such as acidity or fat content. Storage time and contact duration between the packaging material and food likewise affect how much chemical leaches into the contents. Coupled with this, the packaging material and its design also have a significant impact on the extent of chemical transfer onto food items. Therefore, it is imperative to comprehend these determinants when planning strategies for minimising chemical migration and ensuring safety standards for consumable products.

It is crucial to consider the impact of packaging material and design on chemical transfer onto food items. Research has shown that toxic chemicals present in plastic packaging can leach into food and cause severe harm to human health. The use of biopolymer-based packaging presents a potential eco-friendly and nontoxic alternative to synthetic plastics widely used for food packaging because of its ability to reduce contaminants' transfer. Synthetic plastics are usually chosen as a popular food packing material because of their low cost and excellent physical properties; however, they contribute significantly towards environmental pollution by containing hazardous chemicals within them. Given the potential risks associated with using plastic packaging materials, strict regulations must be put in place to govern their manufacture and use.

Alternatives to Plastic Food Wrapping

Plastic food wrapping has become ubiquitous in the food industry because of its convenience and ability to provide a barrier against external factors such as moisture, oxygen, and light. However, concerns have been raised about the potential health risks associated with plastic packaging leaching chemicals that can pose environmental threats and endocrine-disrupting effects. Recent studies suggest that active packaging could offer an effective alternative solution by reducing microbial contamination of fresh produce resulting from water vapor condensation on the inner surface of conventional plastic materials. Emphasising eco-friendly packaging solutions not only benefits public health but also lessens pollution issues while improving shelf life and minimising spoilage rates during storage. Films that are edible and that are derived from other foods are promising candidates for replacing petrochemical-based plastic materials owing to their moderate resistance towards water vapours without causing excessive loss of moisture during storage.

Through our exhaustive research for this book in numerous countries, we have uncovered an interesting pattern of the materials utilised for food preparation and storage. Since the inception of this book, this trend highlights the growing inclination towards alternative packaging techniques that mitigate potential risks of chemical leaching while simultaneously showing their commitment to promoting overall food safety.

As manufacturers and consumers alike become more cognisant of such concerns, there is an emerging demand for products that offer enhanced migration barriers, reduced toxicity levels, and greater sustainability measures. The shift towards environmentally responsible and health-conscious practices will continue to shape the future landscape of the food packaging industry.

Through our extensive research for this book, we found a significant relationship between the frequency of plastic food packaging and the nature of supermarkets. Exclusive and high-end retail stores, in particular, exhibit a more significant inclination to minimise their dependence on plastic food packaging. These niche markets that cater to sophisticated clientele prioritise sustainable and environmentally friendly packing alternatives, and these supermarkets show a greater proclivity to reduce the use of plastic food packaging. This is not an extensive study, but additional studies are required; however, these findings suggest an association between economic status distinctiveness among supermarkets and their adoption of lessened use of plastic packaging practices. Consumers are now expecting corporations to take responsibility for their environmental impact as they become more aware of them. As a result, if supermarkets want to remain competitive while meeting consumer expectations, they must continue to adapt and keep up with changing market trends.

As consumers become more aware of environmental issues, they expect businesses to take responsibility for their impact on the planet. This implies that the economic status of a supermarket plays a significant role in their packaging decisions, as higher-end stores prioritise sustainability because of consumer preferences. Consumers nowadays are becoming more aware of the soaring cost of living and demand to buy food products that are economical yet environmentally friendly without compromising on the product's quality. Though it has been widely acknowledged that glass and ceramic containers are relatively safer than plastic regarding contaminants leaching from materials like plastics, leading to deleterious health effects over time however, some supermarkets still opt for plastic alternatives due to them being cheaper to produce.

As there has been a growing concern among the population about environmental issues worldwide, research efforts are focusing on eco-friendly alternatives to traditional plastic materials. Particularly in the food packaging industry, which depends highly on plastics, biopolymers emerge as an alternative that can mitigate the negative effects of conventional plastic use, enhanced by increasingly strict regulations adopted in various countries against environmental pollution. Instead of using petroleum-based products for making plastics, researchers nowadays turn their attention to bio-based polymers generated from renewable resources such as cornstarch or sugarcane owing to desirable features like biodegradability and reduced chemical migration. Polylactic acid is a popular option because of its ability to decompose naturally after use and has lower leaching characteristics compared with regular plastics. This makes it highly compatible with biological systems, especially in fields that demand this level of compatibility between materials used.

Food packaging plays a crucial role in maintaining food quality and keeping it safe during storage, making it an integral part of implementing sustainable food production systems. Unfortunately, COVID-19 has caused significant changes to global lifestyles that have led to an increase in demand for single-use plastics within the packaging industry. Although this type of plastic helps reduce infection rates by up to 40%, its continued usage contributes significantly as only half (50%) of all waste produced worldwide is plastic, despite low utilisation rates (14). As seen in the image section, in order to prevent contamination with the virus, often fruit and vegetables were wrapped in non-recyclable plastic food wrapping.

To combat this issue and minimise environmental impact when compared with conventional petroleum-derived plastics, there should be more promising alternatives based on biodegradable polymers or materials such as microalgae (a type of seaweed derivative) that can guarantee specific mechanical and barrier properties while being both cost-effective and tailor-made for the solutions that require them. Fortunately, research into these types of alternative solutions has been ongoing for several years before COVID-19's emergence globally; they provide environmentally friendly choices capable of being utilised safely for food contact material because of their chemical safety standards now higher than ever before.

There is a growing demand for better food options that suit dietary needs and ethical issues. In addition to meeting dietary needs and ethical concerns, there is an increasing demand for food packaging that is environmentally friendly. Biodegradable polymers are a promising alternative to traditional plastic materials in the food industry because of their eco-friendliness, and research shows an increased interest in microalgae-based bioplastics as food packaging materials.

However, even though bio-based disposable options are becoming popular amongst consumers during purchasing decisions; these alternatives need careful evaluation since they could still comprise reactive compounds which may lead to allergic reactions or other forms of intolerance issues with some consumers. The safety risks posed by hazardous chemicals used within commonly available wrapping types must also be considered when developing new ways forward for storage produced over time while retaining performance quality, equalling (if not surpassing) traditional low-temperature storage techniques.

Overall, the development of eco-friendly packaging materials is a crucial step towards reducing environmental pollution and meeting consumer demands. As countries adopt more stringent directives against environmental pollution caused by plastics, biopolymers can play a significant role in reducing the impact of conventional plastic materials on the environment across various sectors, such as manufacturing and production industries.

Numerous factors are driving the demand for eco-friendly packaging materials, including consumer preferences for biodegradable and recyclable options. To meet these demands and reduce environmental pollution caused by plastics, the adoption of biopolymers as packaging materials is gaining popularity across various sectors.

One way in which the promotion of sustainability through the reduction of plastic waste can be achieved is by adopting refillable products in supermarkets. Refillable systems provide customers with the opportunity to replenish their supplies by utilising bulk dispensers or refill stations, minimising single-use packaging. For widespread adoption, active cooperation from both consumers and supermarkets is essential. Supermarkets can augment this effort by broadening their range of offerings with more partnering suppliers who offer such options as well by providing incentives like discounts or loyalty programs for users patronising these services. Generating comprehensive knowledge among consumers regarding sustainability advantages and convenience associated with this system plays a significant role in encouraging greater participation in creating a sustainable shopping culture.

During our research for this book, it has become clear that the trend of implementing refillable systems is not limited to a few specialist far shops. In fact, various businesses have already incorporated these eco-friendly practices in their operations and they could serve as examples of wider adoption across the retail industry. The demand for sustainable packaging materials stems from numerous factors, including consumer preferences for recyclable and biodegradable options alongside other driving forces, such as increased environmental awareness and government regulations.

Some progressive retailers like Cobbs at The Farm Stratford-upon-Avon go even further with dedicated refill sections catering to not only foods but also liquid soaps, washing detergents and frozen food products sold by weight or portions, allowing customers to purchase precisely what they need while reducing waste. This innovative approach highlights that embracing circular economy principles can offer both environmentally sound solutions and business advantages through reduced costs associated with packaging disposal over the life cycle of a product.

The demand for eco-friendly packaging materials is on the rise, and businesses are taking note. Many retailers have already adopted sustainable practices such as refillable systems, which not only cater to consumer preferences but also offer cost-saving advantages. The need for recyclable and biodegradable options is just one aspect of the larger initiative towards increased environmental awareness and government regulations that regulate plastic usage. With continued innovation in eco-packaging solutions, the trend will continue towards achieving a more circular economy, reducing waste and improving sustainability in every industry.

The environmental impact of plastic packaging materials on food products is a growing concern because of increasing evidence of microbial contamination and potential health risks associated with toxic chemical leaching. There has been a push for the development and use of biopolymer-based food packaging as an eco-friendly alternative to conventional plastics. Biopolymers are derived from natural sources and can be easily degraded without harming the surrounding ecosystem, thus reducing pollution caused by traditional plastic materials.

However, ensuring that these biodegradable alternatives meet performance requirements and keep satisfactory barrier properties at sub-ambient storage temperatures for their wide-ranging applications in the food packaging field is important. This issue highlights the need for further research into developing effective bio-based materials that not only meet sustainability goals but also provide adequate protection against moisture migration or oxygen permeation while maintaining product freshness during storage.

Transitioning away from petrochemical-derived plastic films presents an opportunity for companies to reduce pollution levels while simultaneously addressing current issues posed towards both human health risks. To summarise, active packaging, eco-friendly solutions and proper material selection are essential aspects to consider when addressing microbial contamination issues associated with plastic wrapping in the food industry.

The current efforts and forthcoming directions to deal with plastic packaging for food involve measures aimed at mitigating environmental impact while endorsing sustainable substitutes.

Scientists are conducting research on biodegradable and degradable materials made from renewable resources like plant polymers such as PLA (polylactic acid), together with cellulose-based wrappings.

Recent advancements in nanotechnology present possibilities for creating smart, active wrapping materials capable of identifying the freshness levels of foods and preventing spoilage. Plastic wrap recycling technologies also continue being refined to enhance recuperation efficacy as well as the re-usability of wraps.

Collaborative partnerships by experts drawn from industries and academic institutions paired up with governments seek initiatives that foster circular economic models, thus minimising ecologic harm brought about by plastics while concurrently innovating better wrapping alternatives designed to ensure safety standards are met through joint effort promotion campaigns geared towards decreasing waste associated. An example of this is BioNet. Agriculture Ltd specialises in edible bioplastics, focusing specifically on the agricultural sectors. This is a collaboration project between Imperial College London and private investors. This company makes edible plastic which is used for cattle bale wrapping, which cattle can then eat and thus eliminating the need for the disposal of plastic waste.

To combat the pressing issue of plastic waste and its damaging effects on the environment, there has been a surge in interest in sustainable packaging made from biopolymers derived from renewable agri-industrial sources. The benefits of this approach are two-fold: it provides an eco-friendly solution while also being cost-effective because of the versatile uses of bioplastic polymers. These include items such as disposable cutlery, food trays, bags, gloves and more durable products like textiles, consumer goods such as razor handles or toys, automobile parts, and even building materials.

Furthermore, Dilkes-Hoffman et al. have discovered that using biodegradable plastics isn't sufficient to eliminate all waste produced by food packaging alone. Proper design considerations regarding how we package our foods are crucial for reducing added wastage; one must always consider environmental factors when choosing which type of wrap you plan on utilising. To help minimise these negative effects, bi-products which would have been considered waste, such as coconut husks, are being reused, creating new life from otherwise discarded substances.

In conclusion, the food packaging industry has recognised the need to replace plastic packages with more environmentally friendly alternatives. With plastics becoming a waste management problem and harming the environment, it is crucial to develop new polymer-based films that are biodegradable, making them commercially valuable as well while replacing petroleum-based plastics that have a short lifespan. Using biopolymers derived from renewable agri-industrial sources has become an attractive solution for sustainable packaging that is cost-effective and versatile in its uses, such as BioNet. Collaboration between industries, academic institutions, and governments are also collectively seeking initiatives towards circular economic models by increasing the recyclability of plastic food wrapping as well as promoting campaigns geared towards reducing waste or using waste bi-products as packaging. This approach will provide long-term benefits to our planet and ensure the sustainable growth of industries while mitigating negative environmental impacts. Considering the environmental problems caused by plastic waste, it is important that the food packaging industry continues to make strides to adopt more eco-friendly alternatives.

Pathogens in food and plastic food wrapping

Plastic-wrapped foods are a common and convenient choice, but they may harbour various pathogens that pose health risks. Bacteria such as Salmonella, Listeria monocytogenes, and Escherichia coli are commonly associated with these types of food items. Research shows that plastic packaging can become contaminated by these pathogens during the production process or through handling practices that do not follow proper hygiene standards. This contamination can occur from raw poultry and eggs, which often contain Salmonella, as well as ready-to-eat foods like deli meats and soft cheeses, which may contain Listeria monocytogenes.

An essential aspect to consider in ensuring the safety of plastic-wrapped foods is proper handling and storage, coupled with adherence to strict hygiene protocols. Although it is not the plastic material that poses a direct threat but rather its mishandling by food workers, consumers may mistakenly believe that packaging alone guarantees the product's cleanliness when, in reality, this can often be far from true. Therefore, care must be taken to prevent possible pathogen contamination resulting from incorrect practices during preparation or consumption of these wrapped products.

Research has shown that raw poultry and eggs could be contaminated with Salmonella, while Listeria monocytogenes bacteria can be found in ready-to-eat foods like deli meats and soft cheeses. These pathogens pose a higher risk of contamination when they come into contact with plastic wrapping materials. It is important to take preventive measures against cross-contamination when handling any commodity wrapped in plastics to avoid exposure to harmful diseases resulting from poor hygienic conditions surrounding the storage or preparation processes.

In order to address concerns regarding plastic food packaging, it is important to take prompt action when potential threats to its recyclability and contamination are identified. Although most plastics utilised for this purpose are recyclable, contamination with Listeria can prove problematic for the recycling process.

Specifically, Listeria monocytogenes have been shown to thrive and multiply on plastic surfaces owned by food packaging companies as well as in delis and homes. Given that such contamination poses a risk of pathogen transmission between materials or products, contaminated packaging may be unsuitable for reuse or recycling purposes, as this would then contaminate not only the original plastic wrapping but the whole bundle of plastic to be recycled.

This is because bacteria such as Listeria monocytogenes thrive on plastic surfaces commonly found in food packaging companies. As reliance on plastics grows in the realm of food distribution networks, waste attributable to single-use wrappers and vessels now exceeds one million tons each year, clarifying that urgent steps towards sustainable alternative sources must be taken without delay. In order to address the growing concerns regarding plastic food packaging, urgent and effective measures must be taken as while most plastics used for this purpose are recyclable, contamination with Listeria and other pathogens can lead to serious problems during recycling it is crucial that all actors in this area can identify and responds appropriately to contaminations as they occur.

Some observe that contaminated packaging poses a significant risk of pathogen transmission between products or materials, resulting in unsuitability for further use or recycling purposes. There have been many initiatives undertaken by different stakeholders towards finding more sustainable alternatives as the dependence on single-use wrappers and vessels continues to increase significantly every year, leading up to over one million tons of waste each year.

Studies carried out by Barro et al. show that there could be numerous factors contributing towards contamination, which includes poor handling practices among food handlers who may not maintain optimal hygiene standards when interacting with these bags, allowing pathogens into them. Despite being an effective means of protecting against spoilage, it is crucial we look into developing eco-friendly options given the dire environmental implications surrounding our extensive reliance on plastic in our world today.

By researching newer techniques like thermoforming polymers (such as polyethylene terephthalates), it is possible to minimise the contact area between chemicals from plastics and foods, ultimately resulting in lesser contamination. The development of biodegradable and compostable materials could offer a more sustainable solution to plastic packaging waste.

Single-Use Plastic and COVID-19

The outbreak of COVID-19 has proven to be a watershed moment in human history, throwing the world into an unprecedented global crisis. The rapid spread of this highly contagious and deadly virus has resulted in multifaceted effects that are both far-reaching and significant. Among these, there have been notable environmental consequences which cannot be ignored. Perhaps one particularly alarming consequence is the surge in single-use plastic waste generated because of using disposable personal protective equipment alongside other medical supplies.

During the COVID-19 pandemic, there was a surge in demand for disposable single use plastic masks, food wrapping and utensils. This amplified requirement incurred substantial financial and ecological costs. From an economic perspective, the widespread application of these products caused inflated costs across different fields. As production expenses rose because of high-demand items, buyers also had to pay exorbitant prices when purchasing them. Also added up is their onetime use feature, prompting frequent buying by individuals or businesses making up recurring cost expenditure.

The significant increase in the production and use of plastic, notably single-use plastics, has led to a corresponding rise in plastic waste with severe environmental consequences. While plastic waste was already a global concern before COVID-19, the issue became even more critical during the pandemic due to single-use polymeric materials such as masks and food packaging. This amplified requirement incurred substantial financial and ecological costs. From an economic perspective, the widespread application of these products caused inflated costs across different fields. As production expenses rose due to high-demand items, buyers also had to pay exorbitant prices when purchasing them.

Also added up is their onetime use feature prompting frequent buying by individuals or businesses, constituting recurring cost expenditure. Meanwhile, from an ecological perspective, the increased usage of disposable single-use plastic has added to existing environmental concerns over waste management.

Rapid manufacturing and disposal of disposable plastics resulted in increased noise pollution and landfill sites while also posing contamination risks for wildlife populations. Humans who consume these contaminants face similar health hazards through ingestion from ecosystems like water bodies, including rivers or oceans where they can accumulate. Various studies suggest that increased usage of plastic-based materials during the pandemic has led to several unintended ecological implications across urban areas worldwide, where effective solid-waste management practices remain challenging at best.

This poses an existential threat to our planet's sustainability over time with cascading effects on ecosystem services such as clean water supply and air quality regulation, among others, all severely affected by these events.

The build-up of plastic waste has harmful effects on natural environments, exacerbating the global crisis of plastic pollution and highlighting the need for improved strategies at both individual and corporate levels to preserve recycling infrastructure. The solution to this crisis includes reducing plastic consumption, improving post-consumer plastic separation and processing, exploring sustainable replacements for disposable products, preventing microplastic leakage, and enhancing recycling rates. The recent COVID-19 pandemic has worsened the situation, as there has been an alarming surge in single-use plastic-based facial masks use, which contributes to the accumulation of plastic waste.

This increase poses detrimental effects on environmental health and biodiversity, as well as human wellbeing, because of the ingestion of microplastics leading to various physical and neurological effects such as endocrine disruption or impairment of kidney and liver functions. It is essential to recognise that this plastic waste crisis affects not only the present but also future generations, and inaction will undoubtedly result in further destruction of the environment.

The COVID-19 pandemic has not only presented a global health crisis, but it also accelerated the accumulation of plastic waste due to increased usage of single-use, plastic-based personal protective equipment like surgical masks. This surge in plastic consumption has led to various environmental challenges and improper management of waste, which could lead to negative impacts on animal and human health. Proper disposal practices for non-biodegradable PPEs, particularly surgical masks, have become imperative as their presence can pose serious implications for our ecosystems.

As reported by the University College London's Plastic Waste Innovation Hub study, continuous use of one disposable face mask per day for an entire year would cause approximately 66,000 tonnes of contaminated plastics generating additional pressure on our ecosystem at large. Regional lockdowns and stricter hygiene practices have further escalated plastic waste generation through the use of everyday single-use plastics like shopping bags, coffee cups, and takeout food containers.

The creation of single-use plastic masks and packaging causes the extraction of non-renewable resources like crude oil and natural gas. This results in further depletion of these already limited resources. In addition, their production entails energy consumption along with greenhouse gases releases, which contributes to climate change concerns. The production of single-use plastic masks and packaging has far-reaching environmental consequences beyond resource depletion. The manufacturing process alone contributes significantly to greenhouse gas emissions, which exacerbates global climate change concerns. This is because the creation of these products usually causes energy-intensive procedures such as extraction, processing, transportation and disposal.

Along with that, it also results in air pollution from burning fossil fuels like crude oil or coal during different stages of their lifecycle, leading to severe health issues for both humans and wildlife alike. Thus, there needs to be a concerted effort towards reducing reliance on non-renewable resources by promoting sustainable alternatives such as reusable cloth masks made of cotton or other eco-friendly materials that are much less harmful in terms of carbon footprint when compared with its counterparts.

Clearly, there is a pressing need for more education campaigns aimed at increasing people's awareness about responsible use of plastics, including how they can assist with proper disposal measures. While curbing inappropriate levels of consumption remains key to combating pollution caused by plastics; research efforts should focus both on identifying practical alternatives for conventional products made from non-degradable materials, such as biodegradable combined with recycling methods designed around existing infrastructure would redress these challenges effectively.

So, the COVID-19 pandemic has undoubtedly revealed and escalated the negative effects of single-use plastics. The drastic increase in plastic use during this time because of personal protective equipment has resulted in improper disposal methods, leading to significant environmental degradation, resource depletion, and human health concerns. It is imperative that we address this issue through comprehensive education campaigns aimed at responsible plastic reduction strategies. Promoting sustainable alternatives like renewable materials for packaging could aid in minimising air, soil and water pollution, which may result from incorrect waste management processes.

Recycling initiatives must be prioritised globally along with governance on production limitations because surgical masks comprise non-biodegradable material that can end up polluting unprotected natural bodies such as oceans. These measures are critical towards creating a future where every individual contributes actively towards lessening their carbon footprint while preserving our planet for generations ahead.

This approach would safeguard nature's well-being from toxic substances having adverse consequences if not channelled properly or unchecked altogether by authorities concerned about protecting lives amid pandemics like COVID-19.

Single-Use Plastic in a Hospital Setting

Using single-use plastics in healthcare settings, such as disposable syringes, gloves, and sterile packaging, has played a crucial role in upholding rigorous hygiene standards and efficient operations. These items have been essential for mitigating the risk of infection and maintaining patient safety. However, there is growing awareness about the environmental ramifications associated with the widespread reliance on these plastics.

Hospitals are now grappling with the issue of plastic waste generation, acknowledging their role as significant contributors to pollution and landfill accumulation. Using single-use plastics in healthcare settings, such as disposable syringes, gloves, and sterile packaging, has played a crucial role in upholding rigorous hygiene standards and efficient operations. These items have been essential for mitigating the risk of infection and maintaining patient safety.

Healthcare institutions need to take a serious look at their operations, recognising the importance of sustainability. They must understand that while maintaining hygiene and patient safety is crucial, they also need to address the environmental impact caused by single-use plastics.

Many healthcare providers in Ireland and in the EU in their commitment to progress, are embarking on a journey of innovation and ethical responsibility regarding their plastic waste. They must carefully explore alternative solutions to plastic waste, focusing strongly on environmentally friendly materials such as biodegradables. They must also implement comprehensive recycling programs that demonstrate their dedication to reducing pollution and addressing the growing issue of plastic waste generated within their facilities.

In addition to their clinical goals, they recognise the importance of incorporating sustainability into all aspects of their operations. An alarming issue they face is the increasing amount of plastic waste generated by catering services, which has been further amplified because of the impact of the COVID-19 pandemic. To tackle this challenge, the organisation is actively exploring alternative options for single-use plastics in their catering services, such as biodegradable or compostable materials, which align with their commitment to minimising environmental harm and promoting a more sustainable future.

In the wake of these efforts, the healthcare institution stands poised at the precipice of transformative change, demonstrating a profound commitment to a sustainable future. Recognising sustainability as a core aspect of their operations not only shows their commitment to protecting the environment but also showcases their ability to adapt and thrive amid unprecedented obstacles.

These healthcare institutions are at a critical juncture, ready to undergo significant changes that reflect their strong commitment to sustainability. By recognising the importance of integrating sustainability into their operations, they not only showcase their dedication to preserving the environment but also show their ability to adapt and thrive in the face of difficult circumstances.

Through actively seeking environmentally friendly options for catering services and incorporating sustainability throughout their operations, these healthcare organisations serve as an exemplary model for the industry. They are not only prioritising the well-being of patients but also showing a commitment to preserving our planet. By adopting this comprehensive approach, they are leading the way towards a future where healthcare institutions go beyond healing to foster a healthier and more sustainable world.

Anne Hayden, MSc

The Toll of Plastic

"Once upon a time, I was falling in love, but now I'm only falling apart."
(Total Eclipse of the Heart, Bonnie Tyler, 1983)

The economic implications of plastic in our society are complex and diverse. Plastic has become an indispensable element in our everyday existence, with its versatility driving its extensive utilisation across various industries. From packaging materials to car parts, to medical supplies, plastic plays a vital role in many sectors, therefore contributing substantially to the economy. Currently, the global production of plastics comprises about 320 million tons per year, with one-third of all produced plastic dedicated to packaging materials (15). This emphasises the close involvement of the food packaging industry in the production of massive amounts of plastics, which generate both economic burdens and ecological effects.

One of the main factors contributing to the extensive utilisation of plastic is its cost-effectiveness. Plastic materials are frequently more affordable to manufacture compared to alternatives like glass or metal. This economic advantage also applies to transportation and storage, making it an attractive choice for businesses seeking cost-minimisation strategies. In addition to its cost-effectiveness, plastic also offers other economic benefits. Plastic is lightweight and malleable, allowing for reduced shipping costs and easier handling during production.

However, the economic drawbacks associated with plastic materials are increasingly apparent. The environmental repercussions stemming from plastic pollution, including the costs of clean-up efforts and damage to ecosystems, are substantial. As a result, there is an increasing demand from both government entities and consumers for more sustainable alternatives. This shift towards sustainability may lead industries heavily reliant on plastics to incur higher production expenses and potentially experience disruptions in their operations.

Our fascination with plastic has taken a significant turn, bringing to mind the lyrics of Bonnie Tyler's "Total Eclipse of the Heart": "Once upon a time I was falling in love, but now I'm only falling apart." The undeniable consequences of plastic pollution and waste are becoming more apparent, and it is crucial that we change our actions. We are, in essence, constructing a towering economic and environmental dilemma that we must confront head-on. This complex issue is not something that can be ignored or pushed aside as it becomes a pressing reality that we must face directly and take action on.

The economic hardships stemming from disposable plastic products disproportionately effect-income individuals, a predicament driven by several interconnected factors. Lower-income communities often find themselves contending with the repercussions of improper disposal of plastic packaging from other, more affluent areas. Individuals in these impoverished communities are burdened by plastic pollution, suffering more severe consequences from clogged drainage systems, increases in vector-borne diseases, and reductions in tourism compared to affluent areas (16).

This glaring disparity underscores the urgent need for fair solutions to address the economic challenges posed by disposable plastics, promoting fairness and sustainability in our society. The widespread use of disposable plastic products can have severe environmental repercussions that undermine people's livelihoods, especially in industries heavily dependent on a sustainable environment. The prevalence of single-use plastics exacerbates these difficulties and adds to the financial burden faced by marginalised communities. Plastic pollution has detrimental effects on key economic sectors such as agriculture, fisheries and tourism (17).

The social toll exacted by the effects of plastic waste is staggering, but it is also economically significant. In 2019 alone, the projected lifetime cost of plastic production stands at a staggering $3.7 trillion (10). This shows the colossal financial ramifications associated with plastic's persistent presence in our environment. What's even more alarming is that our current global strategies for tackling the plastic crisis are faltering. Without swift and decisive action, the projected societal lifetime cost of plastic produced in 2040 could soar to an astounding US$7.1 trillion (18). To put this into perspective, it equates to approximately 85% of the world's health expenditure in 2018 and surpasses the combined gross domestic product (GDP) of economic powerhouses like Germany, Canada, and Australia in 2019. Urgency and innovation are paramount in addressing this mounting economic and environmental challenge (18).

The economic consequences of plastic in our society are significant and diverse. Plastic's widespread use in different industries makes it a vital contributor to our economy, mainly because of its affordability, lightweight property, and adaptability. However, the environmental impact of plastic pollution is becoming more apparent as clean-up expenses and ecological harm continue to rise.

As a result, there is an increasing need for sustainable alternatives that might disrupt industries heavily reliant on plastics. We find ourselves at a pivotal moment, confronted with an immense dilemma that has both economic and environmental implications. The costs associated with plastic production and pollution are extremely high, reaching staggering levels. It is crucial that we take immediate action to prevent a future where the societal consequences of plastic pollution outweigh the healthcare expenditures of entire nations. It is essential to recognise that low-income individuals and marginalised communities bear a disproportionate burden when it comes to the economic impact of plastic, underscoring the urgency of the requirement to find fair solutions to plastic waste for a fair and sustainable future.

CobbsAtTheFarm, Stratford Upon Avon
Refillable Soaps and Shampoos and Refillable Washing Products

Fortnum & Mason London

CONVERSATIONS ON PLASTIC

Irish Supermarket Plastic Wrapping over Apples and Styrofoam Tray

Single Aubergine in Plastic, Lund, Sweden

Straw Made From Pasta, Meloneras, Las Palmas

Vegetables Wrapped in Plastic, Copenhagen, Denmark

Chapter 3
Making a Splash:
Plastics in Our Oceans

"Do you ever feel like a plastic bag drifting through the wind, wanting to start again?"
(Fire Work, Katy Perry, 2010)

Plastic pollution in our oceans is a pressing global issue that brings attention to the significant impact of human activity on fragile ecosystems. From discarded bottles to abandoned fishing gear, synthetic materials have invaded every part of the marine environment, causing extensive damage. As we reflect on Katy Perry's thought-provoking lyrics - "Do you ever feel like a plastic bag drifting through the wind, wanting to start again?" - it becomes clear how millions of tons of plastic waste roam aimlessly, posing a threat to marine life and disturbing the ocean's equilibrium.

Beginning fresh is not a workable choice, considering the significant harm we have caused to our environment. Instead, it is imperative that we take decisive measures and pursue restoration for the pollution we are responsible for. In this chapter, we delve into the severe repercussions of plastic waste on our ecosystems while emphasising the urgent necessity for collaborative endeavours to purify our oceans and revive our marine realm. To combat this crisis, it is essential that we discard any illusions and confront this problem directly with determination; all in order to safeguard the well-being of both our planet and future generations.

The similarities between the contamination of our oceans by plastic waste and the societal acceptance of smoking in the past are quite remarkable. Both situations started with widespread approval and were initially perceived as positive advancements. Plastics, renowned for their exceptional adaptability, offered convenience while cigarettes were once regarded as elegant and even therapeutic.

Over time, the profound impact of these practices has come into sharp focus. Plastic pollution now stands as a significant global threat to marine life and ecosystems, akin to the sobering revelation of smoking's long-term health risks. Our understanding of the environmental toll of plastics and the perils of smoking took years to mature, unmasking the unintended consequences of our choices.

In response, regulations were enacted to address these issues effectively. Smoking witnessed the introduction of warning labels, restrictions on public smoking, and higher taxes, leading to a decline in smoking rates. The magnitude of plastic pollution in our oceans demands urgent, collective action. Just as we successfully altered smoking behaviours through regulation, a worldwide movement to combat plastic pollution is essential, with governments enforcing bans on single-use plastics and individuals adopting sustainable alternatives.

Introduction

The growing rate at which plastic waste is entering our delicate ecosystems has become a global concern. Plastic pollution poses a threat to our environment and ecosystems, specifically the health of our bodies of water and thus on human health. If current trends continue, the United Nations predicts that by 2050, weight for weight, there will be more plastic in the ocean than fish (19).

Addressing this issue is not a simple task and causes a multifaceted approach, incorporating international policy, sustainable corporate practices, and individual acts in everyday life.

Once the plastic becomes waste globally, it has been found that approximately only 10% is actually recycled (20) depending on the year and whose research one reads, and the plastic waste that ends up in our waterways and accounts for 85% of all marine litter (21). This chapter elaborates on various aspects relating to this predicament, such as identifying sources of waterborne plastic contamination along with different types of pollutants while exploring viable alternatives towards its mitigation.

Plastic production year on year is increasing and reached 368 million tons in 2019 (22). Plastic is not a new phenomenon and has existed since 1862 when Alexander Parks first displayed it at the London International Exhibition, describing it as "durable as horn or ivory". However, little did he know the unforeseen effects that this development would pose on our ecosystems and waterways.

While not new in existence, more than half of all plastics ever manufactured have been manufactured since 2000, so the use of plastic has been increasing dramatically; plastic pollution in bodies of water arises from a variety of sources. Traditionally, the primary school of thought among the wider public surrounding the primary sources of plastic and pollution in the waterways was from land-based sources such as improper waste management, insufficient recycling infrastructure, and littering. Plastic products, such as food wrappers, packaging materials, bottles, and bags, are frequently discarded with little thought, resulting in their eventual transmission to waterways via storm drains, rivers, or wind. Land-based plastic is produced by obsolete plastic products, which are swept into rivers by surface runoff, and eventually enter coastal oceans and seas. Once in the water, these plastics can linger for hundreds of years, wreaking havoc on ecosystems.

To address this issue of plastic pollution, both the manufacturing and disposal of plastic waste must be addressed.

Efforts should be directed at producing environmentally acceptable and sustainable plastic alternatives, reducing single-use plastics, boosting recycling rates, supporting responsible waste management techniques, and encouraging consumer awareness and behaviour change.

This requires a multifaceted approach that encompasses international regulations, sustainable corporate practices, and the everyday actions of individuals.

It is critical to identify the origins of waterborne plastic contamination and distinct types of contaminants while examining potential solutions to this problem. Governments and corporations must play a pivotal role in encouraging these initiatives by introducing laws such as extended producer responsibility schemes, plastic fees, and single-use plastic bans. In order to combat the principal sources of plastic pollution in our waterways, we have to tackle the handling of waste and unethical disposal practices.

Encouraging sustainable behaviours in homes and businesses might help to reduce plastic waste pollution. Education and awareness efforts can also play an important role in motivating folks to be responsible for reducing and reusing their waste where possible.

The Scale of the Problem in our Oceans

Traditionally, the public's connotation of pollution from plastic waste is associated with the disposal of plastic waste in landfill with more of an emphasis in recent years on its recyclability but in fact, a large percentage of plastic and plastic food wrapping ends up in our oceans, rivers and lakes. This has evolved into a pressing environmental concern requiring increased public awareness and immediate action crucial.

Plastic contamination in our oceans, rivers, and lakes is a serious environmental problem that should concern all the public. It has been widely acknowledge in the literature and the media that the excessive use of plastic in our day-to-day activities has accumulated in an immense volume of plastic domestic waste and food packaging which is not easily biodegradable and ending up in our waterways (23). While plastic waste in waterways is not a new phenomenon, the understanding of the impact that this has not only on marine life and the environment but the effect that which it has on human health is of growing concern globally (24). There is also an additional disproportionate knock-on effect from this type of pollution on those whose livelihoods rely on these resources for both food and money.

The scale of the plastics issues in our environment is now so large it is estimated that every year that 19 to 23 million metric tons of plastic enter the ecosystem, which equates to approximately 11% of all plastic waste which is generated globally (25). In order to tackle this issue, it is critical to raise public awareness about the negative impact of plastic pollution on our ecosystem and promote the use of eco-friendly alternatives in their daily lives.

Encouraging consumers to make informed decisions in their daily lives about the type of plastic that they purchase not only helps to safeguard our planet but also encourages long-term growth for future generations.

In this regard, policies have, and still play a critical role in promoting the adoption of laws that encourage the reduction of plastic waste and increase recycling rates as part of broader sustainability programmes, which, in turn, have a knock on effect on the ecology and water health of our oceans, rivers and lakes.

Together, through making alternative and more informed choices, we can collectively work towards a more sustainable future with improved water quality, and a reduced carbon footprint.

Plastic contamination in our waterways is immense, and the implications of taking no action are grave. Plastic production has witnessed a remarkable surge over the last few decades, which has consequently resulted in an alarming escalation in plastic waste generation. A considerable portion of this waste finds its way into our oceans, where it presents itself as a major threat to marine life and ocean health. As per current estimations, approximately 8 million tons of plastic enter the oceans annually (26).

The influence of plastic pollution on oceanic ecosystems and their inhabitants is staggering. Marine creatures such as sea turtles, seabirds, and fish frequently confuse plastics with food, leading to severe medical conditions or even death, as discussed in Dr Phil Noone's chapter. The gravity of this situation demands immediate attention from academics and researchers around the globe to find truly workable and sustainable solutions for reducing plastic usage while promoting environmental preservation practices that will benefit us all in the long run! In an ever-evolving world, where the effects of our actions are not only felt presently but also by future generations, it is crucial that we embrace a forward-thinking perspective. This entails going beyond quick solutions and taking into account the lasting consequences of our decisions.

Economically, plastic pollution has significant implications for various industries, including tourism and fisheries. This issue not only affects the aesthetics of coastlines but also disrupts the natural habitats of wildlife and impedes recreational activities like boating or swimming. Plastic waste disrupts natural habitats for marine wildlife, which can lead to severe ecological consequences with long-lasting effects on the health of both humans and other species in affected ecosystems. The interdisciplinary nature of this issue calls for a multifaceted approach from various fields, including ecology, environmental science, economics and policy making, in order to mitigate its harmful effects effectively. It is essential for all sectors to work together, sharing knowledge and expertise to tackle plastic pollution as well as globally, as this is not a singular country issue.

The detrimental effects of plastic waste in our oceans are numerous and complex. Among the most prevalent forms of this pollution are single-use plastics, which include products like bags, straws, and utensils that are designed to be used only once before disposal. These items contribute substantially to the accumulation of plastic debris in marine ecosystems worldwide and can take hundreds of years to degrade fully. Therefore, it is imperative for individuals and organisations to take responsibility and adopt sustainable practices that include reducing unnecessary plastic usage, recycling waste effectively, and disposing of plastics responsibly to curtail the detrimental effects on the oceanic ecosystems.

Straws are one of the most common single-use plastic products found in our waters. In the United States alone, it is estimated that 500 million straws are consumed and disposed of daily in the U.S.A alone (27), perpetuating an already devastating problem. Coupled with this, a 2023 study published in the Journal of Science discovered that up to 8.3 billion plastic straws litter the world's beaches (28). Straws are especially dangerous to marine animals because they are easily eaten and can cause physical harm, like entanglement or choking. Because straws are too small to be easily caught by ocean clean-up operations, they contribute significantly to plastic pollution in our oceans.

Plastic pollution has a serious and negative influence on marine life, threatening the delicate balance of our ocean ecosystems. Marine animals face multiple challenges as debris made of plastic accumulates in our waterways, threatening their existence and disrupting their complex ecosystems. For endangered species such as sea turtles, plastics in our waterways accounts for nearly 50% of these unnatural deaths.

Water pollution in our oceans extends beyond domestic waste, as discarded fishing nets or "ghost nets" (29) also contribute significantly to the problem. These abandoned fishing gears can ensnare various marine species like seals, turtles and dolphins, exacerbating the already existing concerns over plastic debris generated by households. The entangled marine animals face severe consequences, such as restricted mobility, injury and eventual suffocation that frequently lead to fatalities. Such deleterious effects not only impact individual organisms but also pose a threat to their populations and overall ecosystem health.

Another concerning result of plastic pollution is the ingestion of plastic. Plastic waste is frequently consumed by marine species, ranging from tiny organisms to large mammals such as whales in both the forms of visible and invisible microplastics, which will be discussed later in this chapter. The chemical composition of plastics and their affinity for accumulating hazardous wastes pose a grave threat to marine species. The ingestion of plastic waste containing these toxic substances can cause hormonal imbalances, reproductive impairments, weakened immune systems, reduced survival rates and diminished overall fitness. It is imperative for us to address this critical ecological concern holistically through targeted interventions aimed at mitigating all forms of oceanic contamination effectively.

Sources of Plastic Pollution in our Oceans

Plastic production year on year is increasing and reached 368 million tons in 2019 (22). Plastic pollution in bodies of water arises from a variety of causes. Plastic products, such as food wrappers, packaging materials, bottles, and bags, are frequently discarded incorrectly through littering or not using the appropriate waste bins, resulting in their eventual transmission to waterways via storm drains, rivers, or wind. Land-based plastic is produced by obsolete plastic products, is swept into rivers by surface runoff, and eventually enters coastal oceans and seas. Once in the water, these plastics can linger for hundreds of years, wreaking havoc on ecosystems.

The Great Pacific Garbage Patch has been a subject of great concern in the academic and environmental spheres for several years now. It is effectively the result of an accumulation of the waste products gathering in the ocean. Spanning an estimated 1.6 million square kilometres, this massive collection of marine debris located between Hawaii and California is dominated by plastic waste and other non-biodegradable materials that have created what can only be described as a floating garbage island (30).

This disturbing occurrence serves to highlight the devastating effects of human activities on our planet's fragile ecosystems, further underscoring the urgent need for greater global action towards preserving these natural resources. The Great Pacific Garbage Patch, while the largest at an estimated 80,000 tonnes of floating waste, is not the only garbage patch in our oceans and in fact is only one of five significant garbage patches that are growing in size globally (30).

It is important to note that abandoned fishing equipment such as nets contributes significantly to the Great Pacific Garbage Patch in terms of both volume and potential damage. Fishing nets account for approximately half (46%) of the plastic garbage in this area (31). Estimates suggest that there may be as much as approximately 640,000 metric tons of fishing gear at sea worldwide. Coupled with this microplastics account for a further 8% of the total mass of plastics currently in our oceans. As discussed by Dr Phil Noone in more detail, microplastics are small plastic particles that are less than 5mm in size. They have a variety of different sources, including the breakdown of bigger plastic products and the shedding of microfibers from fabrics.

These microplastic particles pose a significant threat to marine life and ecosystems because they can cause when consumed they can cause internal damage and enter the food chain when consumed by marine species. Microplastics can also attract and accumulate hazardous compounds, amplifying their detrimental effects.

To address this issue, there are ongoing activities all over the world to restrict and reduce the production and use of microplastics, which has developed in unique engineering concepts centred on floating devices capable of collecting a vast quantity of plastic waste threats over time have been created by Dutch scientist Boyan Slat in order to help collect and remove plastic from the Great Specific Garbage Patch. Using these technologies is expected to be expanded and have several fleets to remove the plastic and to be financially viable this type of initiative coupled with this at a local level there are steps being taken at a practical level such as reducing the use of single use plastic products.

Coffee cups are a common single-use plastic item that is all-too frequently found in our waterways, contributing to environmental pollution. Disposable coffee cups are used at an alarming rate globally, with the United Kingdom alone consuming approximately 7 million cups each day. As a result, around 30,000 tonnes of paper cup waste are generated annually (32). Unfortunately, these disposable cups pose challenges for recycling because of their waterproof lining made of plastic materials less than 1 in 400 paper cups is currently recycled in the UK (32). As a result, most end up in landfills or oceans where they can take hundreds of years to degrade fully. The plastic liner may also leach toxic substances such as BPA into the environment, causing harm to both wildlife and humans alike if not appropriately disposed of or recycled through appropriate channels.

Single-use plastics like coffee cups have received significant attention globally because of their substantial contribution towards environmental degradation owing mainly due to littering behaviour across various socio-economic groups, and as a result are the top items littered on beaches around the world (33).

As a result, they pose significant threats to aquatic life by increasing the risk of ingestion-related health problems or mortality events among various species who live within these fragile environments. Addressing this issue requires immediate attention from policymakers at all levels, as well as greater public awareness about its negative effects on both environmental health and biodiversity conservation efforts globally.

During our extensive research for this book and frequent travels, we were fortunate enough to witness a plethora of grass-roots movements taking the initiative towards addressing plastic pollution issues. A remarkable example is from our time spent in Italy, where it was customary for restaurants not to offer plastic straws but to provide pasta-made alternatives which decompose naturally without harming the environment. This innovative solution not only eliminated the need for single-use plastic straws but also showcased the creative thinking and adaptability of the local communities.

We encountered many countries where similar initiatives were taking shape. For instance, Ireland has taken significant steps by embracing alternative options like paper or bamboo straws. By shifting away from disposable plastics, these countries aim to mitigate the harmful effects on our planet's health. The efforts to address plastic pollution are not limited to straws alone. We also witnessed an array of ingenious ideas being implemented globally.

For instance, many communities have implemented recycling programs and promoting the use of reusable bags and containers. In some regions, innovative start-ups have emerged, developing sustainable packaging solutions made from biodegradable materials.

These initiatives show the growing awareness and commitment to finding environmentally friendly alternatives to single-use plastics.

Governments and organisations have joined hands to introduce policies and regulations aimed at reducing plastic waste. Some countries have implemented plastic bag bans or imposed levies on single-use plastic items to discourage their use, such as the plastic bag tax in Ireland. Such measures not only contribute to the preservation of the environment but also promote a shift in consumer behaviour and encourage businesses to adopt more sustainable practices.

The fight against plastic pollution is far from over, but these grassroots movements and global initiatives give us hope for a brighter future. As we continue to witness the creative and determined efforts of individuals, communities, and nations, we are reminded of the power of collective action. By embracing innovative solutions, promoting awareness, and making conscious choices in our daily lives, we can contribute to a cleaner and healthier planet for generations to come.

Effects of plastic pollution on the environment - effects on marine life and food chain

The impact of plastics in our oceans is a complex issue that poses significant challenges. The negative effects on the delicate balance of the food chain are widespread and far-reaching, with implications for both marine organisms and humans who depend on seafood as a crucial source of sustenance (34). These factors create intricate interactions between species at all levels, from plankton to apex predators, leading to cascading ecological effects that require immediate action to be taken in order to prevent the further deterioration of this fragile ecosystem.

Microplastics, which are minute particles formed by the disintegration of larger plastic objects, are to blame for the negative effects on zooplankton and other small marine species that consume them (34). As a result, larger species that feed on these afflicted smaller organisms act as vectors, carrying accumulating toxins within plastics up the food chain. Toxin bioaccumulation and bio-magnification endanger predatory fish, marine mammals, and possibly humans at higher trophic levels.

Toxins can disrupt endocrine systems, create reproductive problems, decrease immunological function, and even cause cancer or other long-term health problems (34).

The ecological implications of microplastics are extensive and merit scholarly consideration. Microplastics possess the capacity to accumulate and undergo bio-magnification throughout the food chain following ingestion by smaller organisms. As these creatures progress from being prey to predators, there is an increase in microplastic concentration within their bodies (35).

This means that larger marine animals such as fish, marine mammals, and even human beings may be exposed to higher quantities of microplastics along with associated toxins. The process of bioaccumulation and bio-magnification carries significant potential for disturbing the balance within marine ecosystems, which can have profound effects on biodiversity, population dynamics and overall ecological stability.

Microplastics are a major threat to both marine wildlife and people. Several studies have found that microplastics can be mistaken for food by marine biota, such as bivalves, fish, seabirds, and even turtles. Plasticisers such as nonylphenol and octylphenol leak into these organisms' bodies, as does Bisphenol-A, causing bio-magnification and bioaccumulation at higher trophic levels. Microplastics can cause physical injury, hunger, and even death to marine species because of these processes (36).

Because of their rising abundance in the world's oceans, where they pose significant ecological concerns when consumed by marine species as part of seafood got from aquatic settings or farmed produce, including shellfish.

As a result, there is also the possibility for humans to be inadvertently exposed to increased amounts of microplastics and associated toxins when consuming seafood or agricultural products from marine environments.

Damage to Marine and Costal Ecosystems

The detrimental effects of plastics in our oceans and the knock on effect on the delicate ecosystems balance of our food chain is an interesting area that causes extensive thought and discussion. This topic carries substantial implications not just for marine organisms but also for human beings who heavily depend on seafood as their primary source of sustenance.

The incorporation of plastic pollutants into the food chain poses multifaceted threats, leading to disturbances in energy and nutrient flows within various marine ecosystems. Therefore, there is an urgent need to delve deeper academically into this issue to understand its potential consequences fully and explore viable solutions for sustainable living practices.

Extensive academic research has extensively examined and meticulously documented the detrimental impact of various human activities, such as overfishing, pollution, and climate change, on coral reefs (37).

This multifaceted exploration has not only shed light on the ecological consequences, but also highlighted the social and economic implications for coastal communities heavily dependent on these vital ecosystems. The detrimental effects of plastic pollution on coral reefs, which serve as delicate and vital ecosystems, have been a cause for concern (38). Not only does plastic waste lead to physical harm such as abrasions, lacerations and suffocation in corals, but it also obstructs sunlight that is crucial for photosynthesis - an essential process required by symbiotic algae living within the corals (38).

To add to this already grave issue, plastics often carry harmful toxins and chemicals into reef habitats, causing damage to coral development and reproduction processes. Plastic pollution on coral reefs is a serious concern because of the fragility and importance of these ecosystems.

Plastic pollution on coral reefs can affect the biological dynamics of these fragile ecosystems in addition to the obvious physical and chemical damages. Plastics have the potential to entangle and trap marine species, interrupting their natural activities and potentially causing harm or death. Plastic waste can generate microhabitats that promote the emergence of opportunistic organisms, disrupting the equilibrium of coral reef populations. Plastic decomposition can generate microplastics, which corals and other reef creatures might consume, potentially inflicting internal harm and compromising their overall health.

Plastic pollution's cumulative effects endanger the biodiversity, resilience, and long-term survival of coral reefs worldwide (38).

The negative effects of destroying and losing coral reefs are significant both ecologically and socio-economically. Coral reefs offer important marine ecosystems that facilitate a varied range of underwater organisms, as well as defend coastlines from costal erosion while supporting local economies through fishing and tourism. Thus, the destruction of coral reefs could lead to a significant loss in biodiversity, disturbing the intricate web of life within oceans.

Coral reefs play a significant role in boosting the health and productivity of neighbouring ecosystems, such as seagrass beds and mangroves. These interconnected habitats provide a platform for aquaculture and mariculture activities to take place, including high-value seafood cultivation like fish, shrimp, and oysters. The degradation of coral reefs would injure these interdependent systems, resulting in an adverse effect on associated industries too.

The economic impact caused by damage or loss of coral reefs translates into reduced fishing catch rates, diminished tourism returns alongside increased defencelessness to coastal hazards, exposing other relevant ventures such as aquaculture. The aftermaths are not only limited locally but also nationally etching its effects on livelihoods, food security ultimately affecting economies at large. Preserving with protection is vital not merely from an intrinsic point-of-view but also for bolstering economic growth prospects juxtaposed with supporting resources available to communities coexisting along coastal lines.

Effects on Human Health

The escalation of plastic pollution in our oceans serves as a significant threat to both human health and marine life. Despite the extensive research carried out on the negative impact of plastics on oceanic eco-systems, it is vital to consider its dire involution for human well-being. A major concern arises because of microplastics ingestion by sea creatures, which can cause seafood contamination. When consumed at higher levels within the food chain, these particles ultimately reach humans globally who regard seafood consumption as an essential source of protein in their diets. These minute fragments could incorporate dangerous chemicals such as polychlorinated biphenyls or bisphenol A. Ingestion of such toxins can lead to harmful accumulation within human tissues that may provoke many illnesses like endocrine disruption affecting hormones, sterility or heightened cancer susceptibility among others.

Plastic pollution has far-reaching effects on human health that extend beyond mere seafood consumption. The negative impact of plastic waste extends to ecosystems and food chains, which indirectly impairs human well-being. Our oceans play a crucial role in regulating the Earth's climate and producing oxygen through phytoplankton photosynthesis; however, excessive plastic contamination severely harms marine life while disrupting entire oceanic systems. This disruption leads to dwindling fish populations, escalating biodiversity loss, imbalances within ecosystem functioning with potentially catastrophic consequences for both humans and non-human species alike - including decreasing overall resilience against environmental threats such as disease outbreaks or natural disasters.

Therefore, addressing marine litter is an imperative shared responsibility across diverse levels of governance, from local governing bodies to international organisations – indivisible policy solution will have ripple effects throughout the world at large.

The oceans have a crucial function in lessening the repercussions of climate change by serving as substantial heat reservoirs and absorbing considerable quantities of carbon dioxide emissions from the atmosphere.

As a result, they manage global temperature variations and affect meteorological processes while producing oxygen through photosynthesis using phytoplankton. Nevertheless, plastic waste contamination has detrimental effects on these vital bodies of water despite their crucial roles supporting Earth's environmental balance.

Plastic pollution in our oceans not only harms marine life but also disrupts essential processes that mitigate the effects of climate change. The oceans, essential to the Earth's ecosystem by regulating climate, supporting livelihoods, and providing resources for humans and other organisms, are being significantly affected by human activity.

Greenhouse gas emissions from plastic production pose a serious threat as they contribute to climate change that endangers our planet. Recent studies showed the negative impact of plastic pollution on phytoplankton growth capacity, which diminishes their organic carbon uptake ability, leaving them unable to provide nourishment for marine bacteria or large organisms like blue whales.

Though one of nature's vital CO_2 absorbers - phytoplanktons absorb approximately 50% of atmospheric carbon emissions- sustainability calls for protecting it from anthropogenic activities like the rampant use of plastics that could worsen an already deteriorating situation caused by global warming pressures in years past.

Therefore, it is crucial that we take steps towards reducing plastic waste and mitigating its negative impact on marine life and oceanic processes. We must adopt sustainable practices and invest in research to understand the full extent of plastic pollution's effects on our environment. Immediate action can help reduce greenhouse gas emissions from plastics while also minimising this pollutant entering our oceans.

Let us work together towards a cleaner future for ourselves and generations to come.

International Efforts to Combat Plastic Pollution

Efforts to mitigate the proliferation of plastic waste in oceans across the globe have gained momentum over the past few years. Governments, environmental organisations, and concerned citizens have all realised the critical need to address this issue. The United Nations Environment Programme (UNEP) has been critical in fostering worldwide collaboration and awareness.

A shared understanding among governments. In addition to these efforts, it is important to expand on removing and preventing the issue of plastic pollution in our oceans. Coupled with this examining the impact that plastic waste has on marine life and ecosystems from a scientific perspective, we can better understand how different plastics behave in different environments - both above water and below.

Furthermore, shining a light on potential solutions such as biodegradable alternatives or improved recycling methods through research initiatives will be significant for promoting actionable strategies for addressing oceanic plastic pollution around the world.

Efforts to mitigate the proliferation of plastic waste in oceans across the globe have gained momentum over the past few years. A shared understanding among governments is crucial for the development and implementation of effective solutions to combat plastic pollution. The United Nations

Environment Programme (UNEP) has been critical in fostering worldwide collaboration and awareness. It is not simply the role of one government or country to tackle this issue along and according to the United Nations Environment Programme's 2018 World Environment Day theme, "Beat Plastic Pollution," all countries throughout the world must work together to tackle plastic pollution via concerted efforts.

This shows the importance of international cooperation in tackling environmental challenges.

Governments play an important role in implementing policies that are aimed at reducing plastic waste. However, currently, stringent legal measures, especially targeting marine pollution caused because of plastics, are lacking. One must consider not only international agreements and conventions but also the need for laws, regulations and awareness programmes at both national and global levels. While efforts by scientific communities towards R&D courses have been extensive in removing plastic waste, fulfilling our first aim, however, this waste should not be allowed to enter the oceans to begin with.

At present, the scientific community has been dedicating their efforts toward an R&D course for removing plastic waste and cleaning waterways that have high levels of plastic debris. However, it is necessary to expand academically into greater areas such as having a comprehensive focus on laws, regulations, and environmental awareness programmes associated with effective conversion and mitigation of plastic debris by national and international government organisations. On this note, governments in many developed countries are implementing various policies aimed at reducing plastic waste pollution.

In fact, one form of nature's contribution towards addressing oceanic challenges created by plastics was recently discovered. Researchers from both China and the UK took samples from the Yellow Sea coast –a world heritage site recognised by UNESCO- where they found a "terrestrial plastisphere." This man-made ecological niche's evolution allows ecosystems not only to coexist but also to produce fungi and bacteria that consume these plastics naturally.

Solving problems relating to keeping our oceans free from pollutants will require a collaborative approach involving multiple stakeholders. It is through such collective efforts that we can hope to mitigate the threats posed by plastics in our oceans and prevent further damage to marine life and ecosystems.

With the global threat of plastic pollution in our oceans still very much prevalent, it is essential that we continue to take collective action towards mitigating its lasting impact on marine life and ecosystems.

Significant strides have been made through efforts by various stakeholders such as communities, NGOs, businesses, governments investing in waste management systems and engaging in circular economy practices. However, there is still much work to be done, particularly around educating local communities about the effects of their waste on the environment.

The evolution of a human-made ecological niche offers a glimmer of hope for future solutions to this global challenge. These global efforts show a rising realisation of the critical need to safeguard our seas from the destructive effects of plastic pollution and move toward a cleaner, more sustainable future.

Conclusion

As I bring this section of my book to a close, I reflect on the magnitude of our exploration into the realm of plastic. It has hopefully been for you, as it has for me, an enlightening and thought-provoking expedition that ideally has instilled a sense of duty in each one of us.

As I finish this book from my vantage point at my office desk, gazing out at the lively Galway docks, I find myself back contemplating the significance and the profound symbolism behind those vessels carrying loads of plastic bound for distant shores (in the case of this week's load, it's destined for the Netherlands). The significance of these ships extends beyond their cargo; they represent our collective impact and choices. Throughout these pages, we've delved deep into the intricate tapestry of plastic's influence, from its humble beginnings to its ubiquitous applications, and the sobering reality of plastic pollution.

Plastic, because of its convenience and versatility, has woven itself into the fabric of our lives and often goes unnoticed and taken for granted. Yet, its consequences on our environment, wildlife, and health have become increasingly undeniable.

I have argued throughout this book that the widespread use of plastics has had a hugely detrimental effect on our environment and public health. Most plastic products are designed for single-use convenience, resulting in them being discarded in natural environments such as oceans. This leakage of macroscopic plastics contributes to the formation of microplastics that enter ecosystems through bio-assimilation, posing serious concerns for society's well-being. To address these issues, it is crucial to implement sustainable practices aimed at reducing plastic waste. Companies have already begun taking steps towards this goal by exploring alternatives like bioplastics or recycling oceanic plastic waste. Additionally, supporting renewable materials and bioplastics can help replace traditional plastics while minimising environmental effects and waste generation.

Through our exploration, I hope we understand that plastic is not merely a mundane subject; it is a matter of utmost importance to all of us. It affects the decisions we make, the goods we use, and the lasting impact on future generations. Plastic may not be a glamorous theme, but it is undeniably relevant to our lives, and understanding it is the first step towards change. It presents a complex narrative - a substance that offers convenience but carries significant consequences (much of them negative) for our environment, ecosystems and our own health.

As we conclude this chapter, it is important to remember that our choices as individuals and as a society play a significant role in shaping not only our future but the future for generations to come.

It is essential for us to aim for improvement, to be conscientious about our use of plastic, and to support innovative solutions that can contribute to creating a more sustainable world. We deserve a future where plastic no longer poses environmental risks or health risks but becomes an instrument for positive transformation. The responsibility lies with us, as well as our industry and government, to bring about this change.

I appreciate your participation in this journey, and let us hope that, together, we can create awareness and pave the way for a brighter future regarding plastic consumption.

In closing, I know we all want a cleaner, greener world. The choices about how to achieve this are ours.

GUEST CONTRIBUTORS

The Green Revolution

Forty Shades – A Green Revolution Blooms!

Cathy Fitzgibbon aka "The Culinary Celt"

Introduction

In recent years, the issue of plastic pollution has gained significant attention worldwide, prompting us to seek sustainable and environmentally friendly alternatives in various aspects of our lives. Thankfully, David Attenborough's narration of the harmful effects of discarded plastics on marine wildlife in the BBC's Blue Planet II series cultivated media, public and political attention to plastics pollution[1].

---

[1] Males, J., and Van Aelst, P. (2020). Did Blue Planet Set the Agenda for Plastic Pollution? An Explorative Study on the influence of a Documentary on the Public, Media and Political Agendas. *Environmental Communication*, Vol. 15, pp. 40-54.

Here across the water, on a macro level, the emerald landscape of Ireland, where I call home, attempts to paint a pristine green image internationally (one thinks here of the world-renowned Tourism Ireland branding of the Wild Atlantic Way); the growing global concern of plastic consumption and its environmental effects lurks beneath the surface.

From the microplastics that contaminate our beautiful island and enter the food chain to the packaging used to store and transport food worldwide, the serious issue of plastic pollution poses a genuine threat. Thankfully, because of the mass appetite for embracing sustainable lifestyles, many pro-active Irish communities and businesses are increasingly supportive of the need to safeguard their local ecosystems and beloved land. Following a conference held on the "Food We Buy", inspired by Darina Allen Irish chef, food writer, TV personality and founder of the world-renowned Ballymaloe Cookery School, Killavullen Farmers Market, one of the oldest farmer's markets in Ireland, is a holistic example of the local community working collectively to encourage plastic reduction. Although the market is a commercial outlet for local producers, it also has important educational and social aspects to it through championing "reduce, reuse and recycle" ecologically sound policies. On the business end of the spectrum, Native Events a Dublin-based organisation that specialises in providing sustainable solutions for festivals, events, and arts and culture organisations. This organisation is dedicated to promoting circular economy principles and creating memorable experiences that priorities sustainability, climate action, nature, and biodiversity.

In 2018, my alias, The Culinary Celt, was born out of my love for food, nature, and my desire to share knowledge and fundamental truths about food in harmony with our natural environment resources. Having experienced firsthand the various negative relationships that many people have with food, my work involves helping others to rediscover a mindful appreciation of it through a host of workshops, webinars and a series of writing contributions here in Ireland. These educational pathways feature sustainable, seasonal, zero waste and plastic free approaches. This quote by Wendell Berry—"The soil is a great connector of lives, the source and destination of all"[2] grounds my work and appreciation of food.

This chapter explores the relationship between what I call seasonal eating and the plastic pollution issue here in Ireland, highlighting the challenges food packaging presents along with a variety of practical solutions that we can adopt to live happier, enriched and sustainable lives in harmony with mother nature.

Plastic Perils: The Link Between Food Packaging and Plastic Pollution

Plastic pollution has become a global crisis, causing widespread damage to our oceans, wildlife, and ecosystems. One of the major contributors to this issue is the excessive use of plastic packaging in the food Industry. Single-use plastics, such as food containers, shopping bags and wrappers, are commonly used to preserve and transport food items (as noted in the photographs in this book by my colleague, Conor). Also discussed in Anne and Phil's chapters, it's unfortunate that much of this plastic ends up in landfills or the ocean, taking hundreds of years to decompose, if at all! Many consumers may not fully comprehend the extent of the problem or the alternatives. However, one significant area in which we can

---

[2] Berry, W. (1977). The Unsettling of America: Culture and Agriculture. San Fransisco: Sierra Club Books.

make a substantial impact is improving our fundamental food consumption eating patterns and behaviours.

Ireland, known for its natural beauty and picturesque green landscapes, faces this growing threat from food plastic pollution, as do all countries around the world. The increasing use of single-use plastic packaging in the food industry has led to considerable environmental and health issues that adversely affect local communities. The convenience of plastic containers and packaging has led to a surge in their usage, causing substantial environmental repercussions. Being an island nation, Ireland's reliance on imported food products, often packed in non-recyclable plastic, further intensifies the problem. Producing the highest plastic packaging per head in the EU and having one of the lowest recycling rates, Eurostat 2020 figures show a waste of 61kg per capita in Ireland is nearly double the EU average (34kg per capita)[3]. The Environmental Protection Agency report that in 2021 alone Ireland generated over 200,000 tons of plastic packaging waste, with the food industry being a major contributor[4]. While the country has been making strides in recycling practices, many plastic items used in the food industry remain non-recyclable because of their complex compositions and contaminations, leaving local businesses burdened with the responsibility of managing and disposing of plastic waste themselves. Research findings show the need for educational programmes that promote proper

---

[3]Murray, D. (2023). Ireland's EU topping plastic waste creation needs to be tackled. *Business Post*.

[4]Environmental Protection Agency (EPA). (2021). *Waste Data Release: 2020.*

disposal of plastic waste, which is essential in fostering responsible consumption and waste management practices[5].

The diverse ecosystems of coastal communities across Ireland are another area of society that also suffers directly from plastic litter washing up on nearby shores. Struggling to break down, plastic materials persist in the environment for decades, endangering marine species through entanglement and ingestion[6]. Local fishing communities report encountering plastic debris in their catches, leading to economic losses and potential health risks associated with contaminated seafood[7].

Food packaging and plastic pollution in Ireland pose significant challenges that affect the environment, health, and economy. In relation to health concerns, food plastic pollution raises a vast array of multifaceted issues for local communities. Toxic chemicals from plastics, such as bisphenol-A (BPA) and phthalates, can leach into packaged food and beverages, potentially posing health risks to consumers[8]. Long-term exposure to these harmful substances has been linked to reproductive and developmental issues, hormonal disruptions, along with an increased risk of certain diseases[9]. To address this issue effectively, a multi-faceted approach is required,

---

[5]Kirwan, L., et al. (2017). Public Understanding of Environmental Issues in Ireland: Knowledge, Behaviours and Attitudes. *Environmental Research*. Vol.154, pp. 223-234.

[6]O'Connor, I., et al. (2019). Challenges and solutions to single-use plastics pollution. Marine Policy, 108, 103599.

[7]Gallagher, A.J., et al. (2020). Plastic Pollution in the Marine Environment. A Rapidly Expanding Research Field. Environmental Science and Health. Vol.16, pp. 1-6.

[8]Andra, S.S., et al. (2015). Emerging trends in Environmental Plasticity: Implications for Human Health. Environment International. Vol.83, pp. 126-139.

[9]Schecter, A., et al. (2010). Bisphenol A (BPA) in U.S food. Environmental Science & Technology. Vol.44(24), pp. 9425-9430.

involving policy changes, community engagement, and industry collaboration.

Microplastics in our Food

In addition to the macroscopic plastic waste generated by food packaging, another concerning matter to consider is microplastics in our food. These tiny plastic particles, measuring less than 5 millimetres in size, found in a variety of food products and beverages (also researched and discussed in depth by Phil and Anne) can contribute to a vast number of serious medical issues. Ongoing research suggests the potential health implications of ingesting these microplastics. NUI Galway scientist Dr Liam Morrison, a leading Irish expert on microplastics, has warned about these negative health effects following the publication of research linking microplastics with inflammatory bowel disease (IBD)[10].

Microplastics enter the food chain through several routes. The breakdown of larger plastic items, such as bottles and bags, leads to the formation of microplastics. These particles can contaminate water sources, soil and food, which are then consumed by plants and aquatic organisms, that eventually find their way into rivers and oceans. Then once in the marine environment, these particles ingested by small marine organisms work their way up the food chain. Studies show they can accumulate in the tissues of marine animals, and the growing concern is that they may have adverse effects on

---

[10] Siggins, L. (2022). *Irish Independent Newspaper*, available at https://www.independent.ie/news/chronic-bowel-disease-linked-to-exposure-to-microplastics/41294276.html, accessed at 16:10, 10th September 2023.

human health when consumed through contaminated food and water. Both Phil and Anne have added valuable contributions to this grave issue of concern.

In recent years, multiple investigations have shown evidence of microplastic contamination in food and drink - up to 600 particles of microplastics per kilogram of salt, up to 660 microplastic fibres per kilogram of honey and around 109 microplastic fragments per litre of beer[11]. Seafood also has been found to contain significant levels of microplastics. A further study reported that fish, mussels and oysters harvested for human consumption contained 0.36 - 0.47 particles. Based on average consumption, this study concluded the annual dietary exposure for European shellfish consumers can amount approximately to 11,000 microplastics particles per person, per annum[12]. These reports further highlight the critical role that reducing plastic pollutions plays in maintaining our natural food sources and overall wellbeing.

---

[11] Kuna, A. and Sreedhar, M. (2019). Microplastics in Food Chain. *Health Action*, pp.27-28, available at
https://www.researchgate.net/publication/333719200_Microplastics_in_Food_Chain, accessed at 18:20, 16th August 2023.

[12] Van Cauwenberghe and Janssen (2014). Microplastics in Bivalves Cultured for Human Consumption. Environmental Pollution. Vol.193, pp. 65-70.

Tackling the Food-Plastic Connection: Going Back to our Roots!

"Get back, get back. Get back to where you once belonged."[13]

For contextual analysis, go back in time to our pagan roots. For generations, the rich culinary heritage of Ireland has been deeply ingrained in seasonal eating practices. Our ancient ancestors, the Celts, were attuned to the cyclical rhythms of nature. Their agricultural practices followed the seasons, sowing and harvesting crops in harmony with the Earth's natural patterns. This type of mindful approach to food production ensured the preservation of resources and minimised waste – values that strongly resonate with the current need to curb our plastic consumption.

Then, as Ireland developed into a modern society, the convenience of plastics brought with it a multitude of complex unforeseen consequences. To help tackle this growing national issue, government policies and educational centres of excellence across Ireland have done stellar work in reducing the convenience of using plastic, through decades of reformation and education. A fantastic example of this was the introduction of plastic bag levy by the Irish government in 2002 to help change consumer behaviour and decrease the population's dependency on plastic bag usage. The introduction of 15 cent environmental levy on plastic bags at points of sale dramatically

---

13 Lennon, J. and McCartney, P. (1969). "Get Back" The Beatles and Billy Preston: Apple Records.

reduced the population's plastic consumption and adverse effects it had on Ireland's landscape.[14]

University College Cork builds its national and international reputation as a leading green university. Their 'Green-Campus programme' has seen the university take significant strides forward in its sustainability. In 2010, it was the first third level education institute worldwide to receive the Green Campus award. Subsequently, in 2017, the university banned disposable coffee cups from its library as part of the Green Library campaign and proceeded the following year to launch Ireland's first single-use plastic free café! Further steps were recently undertaken to eliminate all single-use plastic in its operations throughout on campus dining, shops and vending machines from January 2023. This latest initiative by the University, titled 'Plastic Free UCC', focuses on continuous improvement for sustainability across its activities.[15] Implementing these types of plastic-free initiatives has been both practical and educational ways to make a tangible difference and improve the overall sustainability of the campus.

Our modern-day reliance on plastic packaging has escalated exponentially, contributing to the issues in our food chain and the mounting global plastic pollution crisis. So, these types of continuous improvement initiatives are of paramount importance for future generations.

---

[14] Anastasio, M. and Nix, J.(2016). Plastic Bag Levy in Ireland. *Institute for European Environmental Policy* https://ieep.eu/wp-content/uploads/2022/12/IE-Plastic-Bag-Levy-final-1-1.pdf, accessed at 18:40, 12th September 2023.

[15] UCC (2022) – University College Cork Go Plastic Free https://www.ucc.ie/en/news/2022/university-college-cork-to-go-plastic-free.html, accessed at 19:15, 9th September 2023.

CELT by Name, CELT by Nature

Seasonal eating, rooted in our tradition of respecting nature's cycles, aligns with the spirit of Ireland. Contributing to the food education space for the past five years has enabled me to witness first-hand the benefits of seasonal eating and how this simple practice can effectively contribute to reducing plastic waste. The consumption of locally grown, seasonal produce presents a significant opportunity to decrease the demand for plastic packaging and transportation. Out-of-season produce that's often heavily packaged to maintain freshness during long-distance transportation does not serve our wellbeing. In an earlier chapter contribution to a highly informative an educational Mental Health For Millennials book series, I champion the idea that seeking local seasonal foods, to fill our needs in a more environmentally conscious way, affects how we feel both mentally and physically[16].

For the past 15 years, I have adopted a practical, mindful, seasonal eating approach into my daily lifestyle that works in harmony with our natural environment resources. To put it simply, the concept of 'Eating with the Seasons' endorsed in the core ethos of my work promotes a healthy, environmentally conscious lifestyle. Revisiting traditional recipes passed down from generation to generation and embracing seasonal delights can help us reduce our plastic footprint whilst relishing the flavours of heritage and home. The beauty of this movement lies in its simplicity – a return to our roots, honouring nature's cycles, and reducing our impact on the planet. Thankfully, the recent shift back to seasonal eating in Ireland

---

[16] Fitzgibbon, C. (2019). Millennial Culinary Curiosity: Generation Foodie Fuelling Generation Next. *Mental Health For Millennials*. Vol. 3, pp.141-149. Galway: Book Hub Publishing.

has led to a renaissance of support for local farmers and traditional-style food markets. Rediscovering the connection with between food and its sources, allows us to discover the joy of buying fresh, unpackaged vegetables and fruit, whilst reducing our dependency on plastic.

The value of community initiatives to help combat plastic pollution, have been highlighted with the tourism mecca town of Killarney becoming the first town in Ireland to remove all takeaway coffee and teacups – a move aimed at eliminating over one million single-use containers from local waste disposal systems.[17] At the national level, the Irish government also plays a pivotal role in mitigating plastic pollution and fostering a plastic-free culture. Supportive policies, incentives for local farmers and regulations on plastic usage in the food industry will further contribute to a sustainable transformation. This green swap also comes at a cost that many farmers cannot afford. Sometimes, biodegradable plastic can be a lot more expensive than its polyethylene equivalent, so in an ideal world it would be best if farmers could consider ways of reducing their reliance on plastic altogether.

The CELT Mindful Eating Model Approach

On an individual level, mindful eating enables us to make conscious choices about the foods we consume, including their packaging and environmental implications. With a staggering

---

[17] Fleming, R. (2023). Killarney to Become First Irish Town to End Use of Disposable Coffee Cups. *Irish Times* https://www.irishtimes.com/environment/2023/07/11/killarney-to-become-first-irish-town-to-end-use-of-disposable-coffee-cups/ , accessed at 16:25, 18th August 2023.

200+ million plastic cups dumped nationwide in Ireland each year[18]. This makes choosing foods with minimal packaging and avoiding single-use plastics important ways to aid the reduction of our overall plastic waste, whilst promoting more sustainable food consumption practices.

The pillars of my sustainable ethos are displayed in a series of interconnected frameworks and tools: The CELT Mindful Eating Frameworks, Culinary Compass Tool and CELT Mindful Eating Matrix are practical and useful tools to help us consciously make behavioural choices with the foods we consume. In this chapter, I will introduce the latter. The Celt Mindful Eating Matrix binds my other work.

CELT Mindful Eating Matrix

---

[18] Hickey, D. (2023). More Than 200 million Plastic Cups Are Dumped in Ireland each year. *Irish Examiner* https://www.irishexaminer.com/lifestyle/outdoors/arid-41136952.html

Self-Reflection introduces the concept of emotional eating awareness and the secret to gratitude engagement in terms of our relationship with food.

Discovery develops our understanding of eating sustainably, showcasing the benefits of eating seasonally.

Understanding and Acceptance focuses attention on mental shifts to engage the 5 senses (sight, hearing, touch, smell, taste) and their integration into our day-to-day interaction with food.

Self-Love hosts the Culinary Compass Tool, proposing a cross-section of ways to sustain a healthy balanced relationship with food.

To pave the way for a sustainable future, I believe education is key. In recent years, schools and educational institutions in Ireland have become actively responsive, incorporating lessons on the environmental impact of plastic pollution. A series of interactive climate action events and activities are hosted annually around the country. Having taken part in The Make One Change - Cork County Cuts Carbon 2022 event here in the southwest of Ireland, my talk titled 'Eat Sustainably With The Seasons' was hosted to educate attendees on the great power they hold in shaping the future of our natural environment resources with their food consumption lifestyles. Instilling families with this type of valuable knowledge and a sense of responsibility sets the stage for a greener and more conscientious society.

Conclusion

Ireland's journey towards reducing plastic consumption through seasonal eating is testament to the power of ancient wisdom combined with modern consciousness. Aligning with the natural rhythms of the land will facilitate the preservation of biodiversity, helping local communities seize a sustainable future. The Celtic connection to seasonal eating serves as an inspiration to the rest of the world to follow suit and take steps towards healing our planet from the scourge of plastic pollution. As Ireland pioneers the path to seasonal eating, taking strides towards a more sustainable and plastic-free future, the world watches and learns, hoping to emulate this Emerald Isle's green revolution.

In conclusion, seasonal eating is a powerful way to mitigate plastic pollution. This conscious choice not only benefits the environment but also promotes better health, supports local communities and fosters a deeper connection with nature's cycles. As individuals, we have the power to make positive strides by embracing the simple yet profound practice of eating with the seasons.

The Notion of an Ocean

Plastic Use and the Need to Protect our Oceans

Phil Noone PhD Director, Ocean Mindfulness

Introduction

Plastic is everywhere. Due to its design flexibility, versatility, low weight and low cost of production, it is an ideal choice of substance for use in products such as medical supplies, smartphones, disposable cups, food packaging, water bottles, clothing and construction material (Worm et al. 2017). Invented only 110 years ago, plastics are now the most widely used man-made substance.

In 1964, 15 million tons of plastic were produced globally. This grew to 311 million tons in 2014 and further increased to 359 million tons in 2018 (Plastics Europe, 2015, 2019).

Recent evidence indicates that what was originally considered to be a harmless substance is now believed to cause complex toxicity to marine life and to human health via the food chain. In addition, many years of plastic release into the environment has resulted in a major environmental hazard (Cressey 2016), particularly in the ocean where the breakdown of plastic is prolonged (Boucher & Billard 2019), effects on ocean life are severe and options for removal challenging (Gall & Thompson 2015). This chapter explores the impact of plastic on marine life, arguing that we urgently need to reduce our plastic use, reuse and recycle and how new technologies and AI can potentially help this process.

Origins and Development of Plastic

Plastic was first invented by Belgian chemist, Leo Baekeland, who pioneered the first synthetic plastic in 1907. He named this new material 'plastic' after the Greek word *plastikas*, meaning mouldable. His invention combined two chemicals, formaldehyde and phenol, under heat and pressure. This resulted in a resin called Bakelite, which resulted in the growth of a trillion-dollar industry that transformed every aspect of human material consumption (Baekeland 1909). In 1965, the polythene shopping bag was invented by the Swedish company Celloplast. It was designed by engineer Sten Gustaf Thulin and quickly replaced cloth bags in Europe. It is estimated today that the annual production of plastic is almost the combined weight of the human population (7.3 billion people with an average weight of 45 kg) (Worm et al. 2012). This means that we are producing our own weight in plastic every year. Furthermore, if the demand for plastic continues at its current rate, the volume of global plastic waste will grow from 260 million tons per year in 2016 to 460 million tons per year by 2030.

Ocean Health and Pollution

Concerns about the plastic release into the environment were non-existent at first because it was believed the material was not toxic. But in more recent years concerns are growing about the environmental and human hazards of plastic and its impact on the ocean ecosystems, as discussed also by Anne and Cathy (Geyer 2017).

An Oceanographer Eric van Sebille et al. (2015) from the Imperial College London collected data sets that used fine-meshed nets to trawl the ocean to see what plastic they could find. These data sets, containing information from 11,854 individual trawls from every ocean except the Arctic, estimated that in 2014 between 15 trillion and 51 trillion pieces of microplastic were floating in the ocean, estimated total weight of 93,000 tons to 236,000 tons.

Lambeck et al. (2015) researchers in waste management from the University of Georgia in Athens, estimated that 4.8 to 12.7 million tons of plastic ends up in the ocean every year. This is roughly equal to 500 billion plastic bottles every year. But this excludes all the plastic that gets lost or is dumped at sea and all the plastic that is already there. Much is unknown about where plastic ends up or how much of it is in the ocean (Hundertmark et al. 2018).

Types of Plastic Waste

Plastic waste consists of macroplastics, microplastics and nanoplastics.

Macroplastic

Macroplastics enter the ocean in their original size (5-500 mm). The average lifespan of a plastic bag is 15 minutes and yet it can take over 500 years for a plastic bag to break down.

Microplastics

Primary microplastics (1um-5mm) are directly released into the environment in small particles, from toiletries and cosmetics, the erosion of tyres while driving or during the washing of synthetic material, to name but a few. Secondary microplastics break down when exposed to the marine environment, through a process known as photodegradation and the impact of the weather, for example, discarded plastic bottles and fishing nets.

Small microplastics (< 1mm in size) are found in large quantities in the air and water. Their small size, which is invisible to the naked eye, facilitates their ability to be transferred via the food chain and into the human body via seafood. Everaert et al. (2018) estimated a 50% increase in microplastic concentrations by 2100. Boucher & Billard (2019) estimate that tyre abrasion from cars and other motorised vehicles releases an estimated lost rubber of 100mgs/km (1g/10km) for a car (Kole et al. 2017). However, it is estimated that 2-44% enter the marine environment (Wagner et al. 2018).

Nanoplastics

Nanoplastic (< 1um) forms during certain processes, for example, ED printing results in the physical breakdown of microplastics. When these enter the ocean, they can be taken up by oyster larvae and enter the food chain.

Risks to Human Health

People are becoming more aware of the dangers to human health from plastic use. All plastics are made of monomer chemicals that are then combined into synthetic polymers. As mentioned by Anne and Cathy, some of these polymers such as styrene or vinyl chloride are thought to be toxic and carcinogenic. Bisphenol A (BPA) is widely used when producing plastic bottles, and other resins used in food containers are thought to mimic oestrogen hormone-like properties. A laboratory study where Pacific oysters were exposed to microplastics at concentrations similar to that found in ocean sediment where the creatures normally live. Results showed that oysters in plastic-invested water had poorer egg quality and sperm and produced 41% fewer larvae than those in the control group (Sussarellu et al. 2016).

In addition, adipates and phthalates have hormone-like properties and are often added to brittle plastics such as PVC to make them pliable for use in food packaging, toys, and other items in everyday use (Koch 2009). Overall, many of the additives to plastic that are used to make the material soft, hard or colourful are thought to contain carcinogenic or toxicants to human health and to alter our endocrine systems (Kumar 2018). Much research is ongoing into the impact of plastics on human health.

Risks to Marine Life

In addition to dangers to human health, plastic pollution poses grievous harm to marine life, such as entanglement and ingestion of microplastic waste that is thrown in the ocean. All too frequent video clips of sea turtles and dolphins entangled in discarded fishing nets known as 'ghost gear' and other plastic waste, often with fatal consequences. Sea Turtles are thought to mistake floating plastics such as plastic bags for jellyfish causing injury and death. Wilcox et al. (2015) conducted a study in the Arafura Sea, between Australia, Indonesia and Papua New Guinea, which has a large amount of discarded fishing gear, catching between 5000 and 15,000 turtles across 8500 nets examined. Other marine mammals such as dolphins frequently die from entanglement or ingestion or indirectly via uptake of food sources that have ingested nano and microplastics or through ingestion of fauna that has been exposed to chemical pollution. According to Cressey (2016), approximately 90% of seabirds called fulmars in the North Sea who were washed ashore dead had plastic in their gut. What's not so clear and is an ongoing area of research is what exact impact this has on population health.

There is a lot of debate about whether plastic sink and thus accumulate in the deep sea and therefore are not measured by surface sampling (Koelmans et al. 2017) or whether it accumulates in the food chain or swirls around in the ocean. Dai et al. (2018) discovered an abundance of <300um debris increased with depth of water, with artificial fibres being the main plastic type. Whereas, large debris (1mm) decreased with depth and therefore was predominantly found on the surface area of the ocean. Other factors also influence whether the microplastic remains on the surface or sinks to the sea floor: biofouling and water stratification. Biofouling refers to the amount of organisms on submerged surfaces that impact on the buoyancy of plastic (Kooi et al., 2017). Water stratification and circulation, for example, surface and deep water masses, show distinct circulation patterns, but more studies are needed to explore the impact of these circulation patterns on plastic transfer in deep sea areas. Recent public attention has focused on the Great Pacific Garbage Patch as discussed, where plastic collects because of the ocean current called the gyre, but knowledge of this area is in its infancy.

Potential Solutions

There is no simple solution to get rid of plastic pollution. Plastic is a very pervasive material. It is worrying that most of our plastic waste ends up in landfills. Richie & Roser (2017) estimate that in 2015, globally, 50-60% of plastic waste was dumped in landfill, 20-30% was incinerated, and only 10-20% was recycled.

Plastics Europe and the European Association of Plastics Recycling and Recovery Organisations (EPRO) conducted a study of the 28 EU countries plus Norway and Switzerland to monitor plastic waste, recycling and recovery (European Commission 2018; Plastics Europe 2017). This study showed that the collected annual post-consumer plastic waste was 25.1 million tons. This study also highlighted that the plastic material on packaging has the highest recycling rate at 42%. However, Chidepatil et al. (2020) point out that regardless of whether plastic waste goes to landfill, incinerating or recycling the carbon emissions from the production of virgin plastics amounts to 3.8% of the overall global carbon emissions. If this is combined with the disposal and recycling process, then the carbon footprint of plastics is much higher (Zheng & Suh 2019). This points to a very urgent need to take action.

'The Ocean Cleanup'

This project employs a 100 kilometre long floating barrier in the Great Pacific garbage patch with the aim of removing 50% of the surface plastic that resides there. However, some researchers argue that this method of plastic removal from the ocean will disturb fish populations and plankton. Others claim it might be better to position clean-up equipment near the coasts of China and Indonesia, where most of the plastic pollution originates. In 2018, China contributed to approximately 23% of plastic pollution and 21% of waste generated in the world (Plastics Europe, 2019; Ritchie & Roser 2019).

Classify Plastic as a Toxic or Hazardous Substance

According to Seay (2022), we need to classify plastic as a toxic or hazardous substance. This, he argues, might pave the way for how we use and dispose of plastic.

Reduce Plastic Use

Obviously, the first step to reducing plastic waste in the environment is to use less. Before we purchase, ask the question- do we need it? Can we use an alternative substance?

Recycle Plastic

More recycling is certainly a way forward. But this is not a simple solution. Recent studies indicate that over 10,000 additives, processing aids and monomers are used to make plastic, and about 2400 of those are potentially hazardous

(Wiesinger et al., 2021). Sorting out the different plastics is an enormous task. In addition, mechanical recycling shortens polymer chain lengths, with the result that there are only so many times a polymer can be recycled (Seay 2022).

Plastic Circular Economy

A study conducted by Plastics Europe (2019) analysed plastic production between 2011 and 2018. The data showed that 40-48% of plastic are non-recyclable, 19-30% are recyclable and almost 20-27% are complex or unknown categories. Chidepatil et al. (2020) argue that most of the non-recycling plastic has the potential to be converted into waste energy. But at present, most are dumped into oceans and landfills. He and his colleague propose an *AI-assisted segregation system* to recognise, for example, 2 bottles of the same type even if one was misshapen and discoloured. Segregating plastic waste is a costly and challenging task. These researchers have developed robotic multi-sensor based segregation processes using AI that can separate plastic into different categories depending on colour (pigments), chemical composition (plastic type) and sources of plastic waste. In addition, the development of blockchain smart contracts and assists manufacturers to share data, purchase orders and increase the use of recycled plastic.

Other interesting work conducted by Hundertmark et al. (2018) and Wang et al. (2021) suggest that mechanical recycling can be profitable by re-converting waste plastic into biofuels and chemical energy.

The 'Killarney Coffee Cup Project' as briefly by Cathy started in 2023. It reached an agreement between 25 coffee shops, hotels and café to no longer offer single-use disposable coffee cups. By partnering with 2GoCup, it enabled customers to purchase a reusable cup for a 2 Euro deposit, which is refundable when the cup is returned to any of the participating locations in Killarney or any of the 350 existing locations nationwide. It is estimated that this project will remove 18.5 tons of plastic rubbish from local waste centres. And it can easily be replicated throughout Ireland (Ralph, Irish Independent 2023.)

Indeed, Ireland was proactive in its introduction of a 15-cent levy on plastic bags in 2002 with the result that the consumption of plastic bags fell by 90%. The levy not only changed individual behaviour in relation to disposable plastic bag use but it also raised national consciousness about the role each of us can and must play as a collective group to reduce plastic use and achieve a cleaner and more sustainable environment. This levy was increased to 0.22 cents in 2017 and has generated over 200 million euros over 12 years (2002-2013).

Other actions we can take include getting involved in beach/river clean-ups so that we can prevent plastic waste from destroying our beautiful beaches, coastal areas and entering the ocean.

Let us put sustainability first every day from today onwards for a better, clearer environment, better protection of our oceans and marine ecosystems, and better health and wellbeing for all of us.

Conclusion

In 30 years, our oceans will contain more plastic by volume than fish (WEM 2016). If we do not take responsibility for our actions, we are silently destroying and poisoning an entire marine ecosystem (Chidepatil et al. 2020), and damaging human health, most of which remains unknown. We need to reduce the use of plastic, improve waste management and recycle materials to stop them from reaching our rivers and seas.

But plastic will have left its mark. The presence of plastic pollution in the ground, embedded in sediment on the ocean floor and in geological sediments is so extensive that it has become a stratigraphic marker in the Anthropocene. This means that the impact of human activity on the earth's ecosystem can be shown by the examination of plastic pollution, a legacy of the plastic era. According to Eriksen (2014), there will be a layer of rock around the earth that is going to be plastic. Let us take action now to improve the health of our oceans, our marine ecosystems and our own health before it is too late.

References

Anastasia M & Nix J. (2022) Plastic Bag Levy in Ireland. Green Budget Europe. *Institute of European Environmental Policy.* Available at: https://ieep.eu/wp-content/uploads/2022/12/IE-Plastic-Bag-Levy-final-1-1.pdf Accessed on 31/08/2023

Arnaud F. (2019) Value and Limitations of Plastic. *Reinventing Plastics.* 19, 42-43.

Baekeland L. H. (1909) The synthesis, constitution and uses of Bakelite. *Industrial Engineering Chemistry.* 1, 149-161.

Boucher J. & Billard G. (2019) The challenge of measuring plastic pollution. *Reinventing Plastic.* 19, 68-75.

Cressey D. (2016) The Plastic Ocean. *Nature.* 536, 263-265.

De Costs J. P. & Santos P. S. M. & Duarte A. C. & Rocha-Santos T. (2016) Nanoplastics in the environment: sources, fates and effects. *Science Total Environment.* 566, 15-26.

European Commission (2018) *A European Strategy for Plastics in a Circular Economy. Technical Report.* Available at: https://ec.europa.eu/environment/circular-economy.pdf/plastics-strategy-boucher.pdf. Accessed on 20/07/2023

Geyer R. & Jambeck J. R. & Law K. L. (2017) Production, use and fate of all plastics ever made. *Science Advance.* 3(7):e 1700782.

Gall S & Thompson R (2015) The impact of debris on marine life. *Marine Pollution Bulletin.* 92, 170-179.

Hundertmark T. & Mayer M. & McNally C. & Simons T. J. & Witte C. (2018) How Plastics-waste recycling could transform the chemical industry. *Mckinsey on Chemicals, McKinsey & Company.* Available at: https://www.mckinsey.com/~/media/McKinsey/Industries/Chemicals/Our%20Insights/How%20plastics%20waste%20recycling%20could%20transform%20the%20chemical%20industry/How-plastics-waste-recycling-could-transform.pdf Accessed on: 31/08/2023

CONVERSATIONS ON PLASTIC

Jambeck J. R. & Geyer C. & Wilcox T. R. & Siegler M. & Perryman M. & Andrady A. & Narayan R. & Law K. L. (2015) Plastic waste inputs from land into the ocean. *Science.* 347, 623. Available at: https://www.constantinealexander.net/2015/week7/ Accessed on 08/06/2023

Koelmans A.A. & Kooi M. & Law K. L. & van Sebille E. (2017) All is not lost: Deriving a top down mass budget of plastic at sea. *Environmental Research Letters.* 12 (11), 1-9. Available at: https://iopscience.iop.org/article/10.1088/1748-9326/aa9500/pdf Accessed on 10/08/2023

Koch H. M. & Calafat A. M. (2009) Human body burdens of chemicals used in plastic manufacture. Philosophical Transactions. Royal Society of London. *Biological Science.*364 (1526), 2063-2078. Available at: https://www.ncbi.nlm.nih.gov/pmc/articles/PMC2873011/ Accessed on 22/06/2023

Kole M. & Egbert H. & van Nes M.S. & Koelmans A. A. (2017) Ups and downs in the ocean: Effects of biofouling on vertical transport of microplastics. *Environmental Science and Technology.* 51(14, 7963-7971.

Kuhn S. & Rebolledo E.L.B. & van Franeker J. A. (2015) Deleterious Effects of Litter on Marine Life: In *Marine Anthropogenic Litter.* Eds: Bergman U. & Gutow L Klages M. Springer: USA.

Kumar P. (2018) Role of Plastics on Human Health. *The Indian Sources of Pediatrics.* 83, 384-386.

Koelmans A.A. & Kooi M. & Law K. L. & van Sebille E. (2017) All is not lost: Deriving a top down mass budget of plastic at sea. *Environmental Research Letters.* 12 (11), 1-9. Available at: https://iopscience.iop.org/article/10.1088/1748-9326/aa9500/pdf Accessed on 10/08/2023

Plastics Europe (2018) *Plastics: The Facts 2018. Technical Report.* Available at: https://www.plasticseurope.org/download_file/force/2367/180 Accessed on 02/08/2023

Plastics Europe (2015) *Plastics: The Facts 2015. Technical Report.* Available at: https://www.plasticseurope.org/download_file/force/835/180 Accessed on 02/08/2023

Richie H. & Roser M. (2019) *Plastic Pollution. Technical Report.* Available at: https://ourowrldindate.org/co2-and-other-greenhouse-gas-emissions Accessed on 02/08/2023

Repak (2018) *Plastic Packaging Recycling Strategy 2018-2023.* Available at: https://repak.ie/images/uploads/downloads/Plastic_Packaging_Recycling_Strategy_2018-2030.pdf Accessed on 31/08/2023

Sargent N (2018) Plastic bottles found at over 80 per cent of coastal sites: New study findings. *Green News.ie* Available at: https://greennews.ie/plastic-bottles-80-per-cent-coastal-sites-study/ Accessed on 31/08/2023

Seay J. R. (2022) The global plastic waste challenge and how we can address it. *Clean Technologies and Environmental Policy.* 24, 729-730.

Van Debille E. & et al. (2015) *Environmental Research Letters.* 10, 124006.

Wagner S. & Huffer T. & Klockner P, & S. & Wihrhahn M. & Hofmann T. & Reemtsma T. (2018) The Wear particles in the aquatic environment: A review on generation, analysis, occurrence, fate and effects. *Water Research.* 139, 83-100.

Wang Z. & Burra K. G. & Lei T. & Gupta A. K. (2021) Co pyrolysis of waste plastic and solid biomass for synergistic production of biofuels and chemicals: A Review. *Progress in Energy and Combustion Science.* 84, 110899. Available at: https://www.sciencedirect.com/science/article/abs/pii/S036012852030109X Accessed on 01/09/2023

Wiesinger H. & Wang Z. & Hellweg S (2021) Deep dive into plastic monomers, additives, and processing aids. *Environmental Science and Technology.* 55 (13), 9339-9351.

Zheng J. & Suh S. (2019) Strategies to reduce the global carbon footprint of plastics. *Nature Climate Change.* 9, 374-378.

The Future is in Our Hands - A Voice From Gen Z

The Global Plastic Waste Crisis Through the Lens of Gen Z: Exploring Perspectives and Innovating Solutions

Conor MacGiolla Bhuí, Governance, ShadowScript Ghostwriters

Introduction

Generation Z refers to individuals born roughly between 1997 and 2012 who grew up in a world dominated by technology and faced with mounting ecological issues. Our generation has witnessed firsthand the consequences of widespread plastic consumption and disposal practices that prioritise convenience over sustainability.

We have been exposed to distressing images documenting vast patches of floating plastic debris in our oceans or wildlife entangled in discarded packaging materials. These experiences have shaped our perception of environmental responsibility and fuelled our determination to address these pressing issues head-on.

Generation Z, the cohort born between the mid-1990s and early 2010s, has emerged as a pivotal force in shaping consumer behaviour and raising awareness about environmental issues. As Courtney (2020) highlights, this generation exhibits a heightened level of environmental consciousness, which significantly influences their purchasing decisions. Moreover, Liu and Hei (2021) emphasise that Gen Z consumers tend to prefer products from brands with distinct brand identities that incorporate practicality and sustainability into their designs.

As the world grapples with plastic waste, Generation Z (my generation) has emerged as a crucial voice in tackling this global issue. This chapter delves into what I believe to be our unique perspective, shedding light on how we perceive the magnitude of plastic waste and its environmental consequences. I'm fortunate in that I now have two countries to compare approaches; Ireland (where I've lived all my life to date and just completed a degree) and Sweden (where I've embarked on postgraduate studies during the summer of 2023). Both countries have developed distinct strategies and political systems to address issues surrounding plastic waste, which is indeed a significant problem

Here's what we're told! Bursting with youthful energy and innovative ideas, Gen Z offers fresh insights that challenge traditional approaches from other age cohort demographics to address this crisis. Whilst this may or may not be accurate, I think it's fair to say that, as a collective, we have deep concerns about the future of our planet, and are actively exploring motivations for action and solutions that align with our collective values. There is no other way. From reducing single-use plastics to promoting circular economies, Gen Zers all over the globe champion sustainable practices while demanding government legislation to curb plastic production.

Our Concerns

In recent years, concerns about plastic waste have escalated as its deeply negative impact on ecosystems becomes increasingly evident. Plastic pollution poses a significant threat to our planet's delicate balance, affecting marine life, terrestrial habitats, and even human health. As members of Generation Z come of age in this era defined by environmental challenges, we are emerging as influential voices advocating for change.

Unlike previous generations, who largely relied on traditional media channels for information dissemination, GenZers tend to utilise social media platforms as powerful tools for raising awareness regarding critical matters such as plastic waste management.

Through engaging content creation and viral campaigns driven by increasingly popular hashtags like #PlasticFreeFuture or #BeatPlasticPollution, young activists mobilise thousands – if not millions – around shared concerns for our planet's future. As I'm writing this chapter on a GoBus heading from Galway city to Dublin Airport where I will shortly board a flight to Sweden, I must give the example of the 'face' of Gen-Z climate activists, Greta Thunberg. She has her fans and her critics, but her central place in the sustainability debate is not in doubt.

Education plays a vital role in shaping Generation Z's perspective on plastic waste around the globe. With access to vast amounts of information at our fingertips through search engines and online resources from reputable organisations committed to sustainability efforts such as Greenpeace or The Ocean Cleanup Foundation - GenZers are well-informed about the severe consequences associated with excessive plastics usage. Armed with knowledge about microplastics infiltrating food chains as discussed by my colleagues, Phil Noone and Cathy Fitzgibbon in this book, or long-lasting impacts resulting from improper disposal methods like landfilling or incineration, we become catalysts for change within our communities.

A Heightened Level of Environmental Consciousness

One pressing issue at the forefront of Generation Z's concerns is plastic waste around the globe. Eriksen et al. (2023) shed light on the alarming escalation of plastic pollution in our oceans' surface layer. Their study reveals astounding statistics: an estimated global abundance ranging from 1.1 to 4.9 million tonnes of plastic particles weighing less than five millimetres, equating to over 170 trillion particles floating within our marine ecosystems. The gravity of this "plastic smog" demands urgent international policy interventions as emphasised by Eriksen et al. (2023). The detrimental effects on marine life are evident; however, it is crucial to comprehend how Generation Z perceives this crisis and what solutions they advocate for. Furthermore, individual action holds immense significance in tackling this multifaceted problem effectively. While governmental policies play a significant role in curbing plastic pollution worldwide – as will be discussed later in this paper – GenZers acknowledge the importance of personal responsibility. Our generation actively engages in plastic waste reduction efforts, such as participating in beach clean-ups, advocating for sustainable alternatives, and promoting eco-conscious lifestyles among their peers. However, individual action alone cannot halt the global plastic waste crisis. It necessitates a comprehensive policy framework that addresses production, consumption, and disposal practices on both regional and international levels. Policy changes should encompass measures like implementing stricter regulations on single-use plastics or incentivising businesses to adopt more sustainable packaging options.

Ireland and Sweden: A Brief Comparison

My first observation is that Sweden seems more genuinely concerned than Ireland. Through ambitious recycling initiatives, stringent regulations on single-use plastics, and proactive promotion of sustainable alternatives, Sweden has set a commendable example for other nations to follow. At its core, Sweden's response rests upon an intricate web of recycling programmes aimed at diverting plastic waste from landfills. To compare, in 2020, Ireland was the biggest producer of plastic packaging waste per capita in the European Union. The average Irish citizen produced 61.52kg per person. In that same year, Hungary was the second highest producer with 47kg per person. Still, this is a huge difference of nearly 15kg between first and second place. In this same year Sweden, along with their Nordic brothers Denmark and Finland were the leaders of sustainable development in the EU (Kiseľáková, et al., 2020).

In Sweden, initiatives encompass everything from household recycling bins to specialised collection centres for hazardous materials like electronic waste or batteries. To add, when purchasing a drink/liquid in a can (e.g a can of coke), consumers pay the equivalent of an additional five to ten cents. When that consumer then returns the can to a recycling machine (often located in supermarkets), they are given this money back to spend in store. Thus, encouraging correct recycling procedures. Moreover, by implementing efficient sorting systems and investing in advanced recycling technologies such as chemical depolymerisation or mechanical reprocessing techniques, Sweden ensures that more plastic is given a second life rather than polluting fragile ecosystems.

Furthermore, it is worth highlighting Sweden's success in reducing consumption through strict regulations on single-use plastics. The country recognises that prevention is key; therefore, they have implemented policies banning or heavily taxing items like plastic bags and straws while actively encouraging businesses to adopt sustainable alternatives instead. This multifaceted approach not only reduces demand but also fosters innovation by creating market opportunities for eco-friendly products.

In addition to curbing plastic usage through regulation, Sweden actively promotes sustainable alternatives across various sectors. From supporting research and development projects focused on bio-based materials to incentivising industries towards adopting circular economy principles – where products are designed with recyclability in mind – their efforts span far beyond just mitigating immediate consequences.

By championing these measures both domestically and internationally, Sweden is showcasing its dedication towards long-term solutions that address the root causes behind our reliance on plastics.

CONVERSATIONS ON PLASTIC

What appeals to me is that the focus in Sweden lies not just on eliminating harmful products but also on introducing innovative substitutes made from biodegradable materials or renewable resources like plant-based polymers or compostable packaging options.

Moreover, collaboration between government agencies, private enterprises, research institutions, and civil society groups has been instrumental in driving change at multiple levels within Swedish society. Through partnerships with businesses operating along different points of the supply chain - from the product design stage all the way through production processes - Sweden has fostered innovation and the development of sustainable alternatives – and it shows on a daily basis walking around Swedish towns and cities.

Indeed, my own campus at Lund University is a case in point both in terms of research and courses but also on campus sustainable initiatives. For instance, in the 2023 QS Sustainability Rankings, which focused on social and environmental sustainability performance in higher education institutions, Lund University ranked 12th in the world (QS, 2023).

Conclusion

As evidenced by the research conducted throughout this essay, plastic waste poses a severe threat to our environment, ecosystems, and future generations. It is incumbent upon us as members of Generation Z to actively participate in finding sustainable solutions. Throughout this chapter, I have explored various aspects of plastic waste management, including its detrimental effects on marine life and biodiversity. The statistics presented have underscored the severity of the issue and emphasised that immediate measures must be taken to reverse these alarming trends. Additionally, I note numerous potential solutions, such as promoting recycling programmes at local levels, implementing stricter regulations on single-use plastics production and consumption, fostering innovation in biodegradable alternatives, and raising awareness through education campaigns. These strategies offer promising avenues towards mitigating plastic waste pollution globally.

As young activists committed to preserving our planet for future generations, it is crucial that we reflect on our individual roles in combating plastic waste.

By making conscious choices like reducing personal consumption of single-use plastics or advocating for change within our communities and governments alike, we can contribute significantly to a more sustainable future. In contemplating the implications discussed above regarding plastic waste around the globe and potential solutions thereof; one cannot help but recognise that while progress has been made towards addressing this critical issue, there remains much work ahead if we are genuinely devoted to achieving lasting change.

CONVERSATIONS ON PLASTIC

We must continue pushing boundaries by leveraging technology advancements and engaging stakeholders from diverse backgrounds in collaborative efforts aimed at creating innovative approaches tailored specifically toward tackling this complex problem. To conclude definitively; embracing a holistic approach encompassing policy changes coupled with individual responsibility will pave the way forward toward a cleaner planet free from pervasive plastic pollution. Our actions today will shape tomorrow's reality – let us unite as Gen Zers across borders – guided by passion fueled by knowledge – igniting transformative change that ensures an environmentally sustainable world for generations to come.

References

Alves, Bruna, (2023). 'Generation of plastic packaging waste per capita in the EU-27 2020, by country'. Energy and Environment. Waste Management. Statista Research Department.

Courtney, Dalton Alexis, (2020), "Exploring Generation Z'S Envornomental Concerns and its Effects on their Purchasing Behaviours https://digital.library.txstate.edu/bitstream/handle/10877/12102/Courtney-Thesis.pdf?sequence=1

Kiseľáková, D., Steć, M., Grzebyk, M., & Šofranková, B. (2020). A Multidimensional Evaluation of the Sustainable Development of European Union Countries – An Empirical Study. *Journal of Cryptology*, 12, 56-73. https://doi.org/10.7441/JOC.2020.04.04.

Liu, Younan & Hei, Ye, 2021, "Exploring Generation Z consumers' attitudes towards sustainable fashion and marketing activities regarding sustainable fashion". https://www.diva-portal.org/smash/get/diva2:1561061/FULLTEXT01.pdf

Eriksen, M., Cowger, W., Erdle, L. M., Coffin, S., Villarrubia-Go′mez, P., Moore, C. J., et al. (2023), "A growing plastic smog, now estimated to be over 170 trillion plastic particles afloat in the world's oceans—Urgent solutions required", PLoS ONE. https://journals.plos.org/plosone/article/file?id=10.1371/journal.pone.0281596&type=printable

QS World University Rankings: Sustainability 2023

Other Sources

Conlon, K., 2023, "Emerging Transformations in Material Use and Waste Practices in the Global South: Plastic-Free and Zero Waste in India", Urban Science. https://pdxscholar.library.pdx.edu/cgi/viewcontent.cgi?article=1367&context=usp_fac

Policy Department for Citizens' Rights and Constitutional Affairs, "The environmental impacts of plastics and micro-plastics use, waste and pollution: EU and national measures", Directorate-General for Internal Policies.
https://www.europarl.europa.eu/RegData/etudes/STUD/2020/658279/IPOL_STU(2020)658279_EN.pdf

About the Author

Anne Hayden is currently completing her doctoral studies in agricultural economics at University College Dublin, Ireland. She holds a first-class Master of Science from the prestigious UCD Michael Smurfit Graduate Business School in Food Business Strategy and a first-class Bachelor in Agricultural Science. Her research interests include the environmental, social and economic regional impacts of possible policy changes to the common agricultural policy. Anne has been a guest contributor to Mental Health for Millennials Volume 5 (2021) on the theme of resilience in Irish agriculture and Mental Health for Millennials Volume 6 (2022) on hope for women in Irish agriculture as well as to Mental Health for Millennials Volume 7 (2023) on 'being present' in the moment in Irish agriculture. She has also been featured as a contributor in the published college studies series on 'an appraisal of attitudes to the environmental and chemical hazards of plastic food packaging.' Anne is a contributor to the book, Essays on Covid-19 (2023) where her chapter focused on 'finding the joy of exercise during Covid-19'. Anne is a consultant with the Dissertation Doctor's Clinic since 2022 and is Head of Sustainability with Book Hub Publishing since 2023.

Niall MacGiolla Bhuí, PhD is Director of ShadowScript Ghostwriters and the Dissertation Doctor's Clinic with offices in Ireland and Sweden. He has travelled extensively lecturing and presenting workshops across Ireland, Northern Ireland, England, Sweden and coast to coast Canada. Niall is series co-editor, along with Dr. Phil Noone, of the seven book series, Mental Health For Millennials.

He is founder and editor of the #ExploringConnectedness book series. He has authored and co-authored several books with colleagues across various mental health and humanities themes. Niall currently ghostwrites for a range of national and international clients and mentors postgraduates across the university sector. His second book of creative writing will be published in 2024.

Phil Noone, PhD is Director of Ocean Mindfulness. She is a Mindfulness Coach/Lecturer with a nursing background who has travelled and worked extensively abroad in a variety of health and well-being settings. She holds an MSc in Mindfulness Based Interventions, a Diploma in Mindfulness and Positive Psychology, a PhD in Sociology and an MSc in Health Promotion. Phil has lectured for many years at the School of Nursing and Midwifery, University of Galway, and has presented at conferences in South America, Holland, Australia and Ireland.

Phil is series co-editor and chapter contributor along with Dr Niall MacGiolla Bhuí of the Mental Health for Millennial Series (1-7), has published papers and conducted research on the themes of 'home', 'well-being', 'rural ageing', 'resilience', 'environmental action and sustainability' and 'Mindfulness'.

Cathy Fitzgibbon, aka The Culinary Celt is a sales and marketing professional in the media industry for the past three decades. She is passionate about exploring contemporary culinary experiences and actively promotes the areas of food sustainably and food tourism through her food writing contributions and marketing activities under her alias 'The Culinary Celt', using an ethically based farm to fork ethos.

She has also contributed to several Book Hub Publishing books and TheDocCheck.Com publications in addition to presenting academic research papers both nationally and internationally. Cathy consults with The Dissertation Doctor's clinic and most recently published her debut book, 'Eat With The Seasons' which has gone into reprint.

Conor MacGiolla Bhuí is a MSc postgraduate at Lund University in Sweden. He holds an honours Bachelor's degree in English Literature and Sociological and Political Science from University of Galway. Conor has published chapter contributions in two books to date with his own co-authored book, 'Threads of Trauma' scheduled for publication in 2024.